Selected Papers from CUBANNI 2017—"The Fourth International Workshop of Neuroimmunology"

Selected Papers from CUBANNI 2017—"The Fourth International Workshop of Neuroimmunology"

Special Issue Editors

Maria de los Angeles Robinson Agramonte
Carlos Alberto Gonçalves
Dario Siniscalco

MDPI • Basel • Beijing • Wuhan • Barcelona • Belgrade

MDPI

Special Issue Editors
Maria de los Angeles Robinson Agramonte
International Center for Neurological Restoration

Carlos Alberto Gonçalves
Federal University of Rio Grande do Sul
Brasil

Dario Siniscalco
University of Campania
Italy

Editorial Office
MDPI
St. Alban-Anlage 66
4052 Basel, Switzerland

This is a reprint of articles from the Special Issue published online in the open access journal *Behavioral Sciences* (ISSN 2076-328X) from 2017 to 2018 (available at: https://www.mdpi.com/journal/behavsci/special_issues/CUBANNI_2017?view=compact&listby=type)

For citation purposes, cite each article independently as indicated on the article page online and as indicated below:

LastName, A.A.; LastName, B.B.; LastName, C.C. Article Title. *Journal Name* **Year**, *Article Number*, Page Range.

ISBN 978-3-03897-487-1 (Pbk)
ISBN 978-3-03897-488-8 (PDF)

Contents

About the Special Issue Editors

Maria de los Angeles Robinson Agramonte, MD, PhD, is a full professor of neuroimmunology at the International Center for Neurological Restoration, Havana University, Cuba. She earned her PhD (2010) from Havana University in neuroimmunology. Dr. Robinson has kept a sustained program of research in neuroimmunology, neurodegenerative disorders, and neurodevelopmental disorders. Her work has resulted in various award-winning book chapters (2013, 2017, and 2018), over sixty professional articles in national/international journals, book chapters, and one book. Her book "Translational Approaches to Autism Spectrum Disorder" resulted "Reward Immunology 2017" in 2017 and from the National Council of Health Scientific Society in 2018. Her chapter Intech 2013 was also awarded Article of the Year from the Cuban Society of Immunology. She is a preeminent expert on neuroimmunology, and is at the forefront of autism spectrum disorders research.

Carlos Alberto Gonçalves, MD, PhD in biochemistry. Carlos is now a full professor at the Federal University of Rio Grande do Sul (Brazil), working on post-graduation biochemistry and neuroscience programs. Carlos teaches classes on medical biochemistry (graduate studies), protein chemistry, and neurochemistry (postgraduate studies). Carlos is an honorary member of the Cuban Society of Neuroimmunology. He is a fellow researcher 1A from the National Council for Development of Science and Technology (CNPq), working on basic and applied neurochemistry focusing on: activity, expression, and secretion of S100B, a calcium-binding protein; markers of glial activation in neurodegenerative diseases, particularly Alzheimer's disease; and protein phosphorylation in the central nervous system. He has over sixty professional articles in national/international journals, as well as several book chapters.

Dario Siniscalco is biochemist for the Department of Experimental Medicine of the University of Campania, Italy. He earned his PhD in pharmacological sciences from the Second University of Naples in 2004, completed a neuropathology fellowship at the University of Alabama at Birmingham, USA before joining the Second University of Naples staff in 2006. He is a registered member of the various scientific societies, and is one of the founders of the Stem Cell Research Italy association. In 2010, he founded a research group to study cellular and molecular changes in autism spectrum disorders. He is an author and co-author of 90 scientific peer-reviewed papers and 10 book chapters. Dr. Siniscalco's works have received more than 2100 citations so far (h-index 28). He has presented 95 communications/abstracts to national and international conferences. He serves as an Editor in Chief, Editorial Board Member, and Reviewer of more than 75 international journals. His full publication list can be found here: http://orcid.org/0000-0002-3779-2596.

Preface to Selected Papers from CUBANNI 2017—"The Fourth International Workshop of Neuroimmunology"

This Special Issue "Selected Papers from CUBANNI 2017—'The Fourth International Workshop of Neuroimmunology'" will emphasize the important role of neurobiological processes in the extensive field of neurology and behavioral disorders, from a translational approach. It will also highlight the significant work of researchers that have been studying new tools of intervention for these very complex diseases, thus bringing academic research toward clinical assistance. It is important that all of this valuable information is not left behind at the meeting, but rather must be published as a major critical discussion in the scientific community, enabling the real and effective communication of scientific information that drives rapid advancement in the field. This book reviews molecular mechanisms involving the neuropathology of several neurodegenerative and neurodevelopmental diseases, as well as medical advances in the treatment of diseases, working toward the goal of the experts which have written articles for this Special Issue with the intention of increasing scientific exchange and triggering new research focusing on neurobiological mechanisms, as well as paraclinical and clinical approaches regarding diseases and interventions from all participants' perspectives.

Maria de los Angeles Robinson Agramonte, Carlos Alberto Gonçalves, Dario Siniscalco
Special Issue Editors

![behavioral sciences logo]

Conference Report

Neuroimmunology Research. A Report from the Cuban Network of Neuroimmunology

María de los Angeles Robinson-Agramonte [1,*], **Lourdes Lorigados Pedre** [1] and **Orlando Ramón Serrano-Barrera** [2]

[1] Neuroimmunology Laboratory, Immunochemical Department, International Center for Neurological Restoration, Ave 25 # 15805 b/w 158 and 160, Playa, Havana 11300, Cuba; lourdesl@neuro.ciren.cu
[2] Las Tunas General Hospital, Post Office Box 27, Las Tunas 75100, Cuba; orlandosb@infomed.sld.cu
* Correspondence: maria.robinson@infomed.sld.cu; Tel.: +537-2716-385

Received: 7 March 2018; Accepted: 25 April 2018; Published: 8 May 2018

Abstract: Neuroimmunology can be traced back to the XIX century through the descriptions of some of the disease's models (e.g., multiple sclerosis and Guillain Barret syndrome, amongst others). The diagnostic tools are based in the cerebrospinal fluid (CSF) analysis developed by Quincke or in the development of neuroimmunotherapy with the earlier expression in Pasteur's vaccine for rabies. Nevertheless, this field, which began to become delineated as an independent research area in the 1940s, has evolved as an innovative and integrative field at the shared edges of neurosciences, immunology, and related clinical and research areas, which are currently becoming a major concern for neuroscience and indeed for all of the scientific community linked to it. The workshop focused on several topics: (1) the molecular mechanisms of immunoregulation in health and neurological diseases, (like multiple sclerosis, autism, ataxias, epilepsy, Alzheimer and Parkinson's disease); (2) the use of animal models for neurodegenerative diseases (ataxia, fronto-temporal dementia/amyotrophic lateral sclerosis, ataxia-telangiectasia); (3) the results of new interventional technologies in neurology, with a special interest in the implementation of surgical techniques and the management of drug-resistant temporal lobe epilepsy; (4) the use of non-invasive brain stimulation in neurodevelopmental disorders; as well as (5) the efficacy of neuroprotective molecules in neurodegenerative diseases. This paper summarizes the highlights of the symposium.

Keywords: neuroimmunology; neurodevelopmental disorders; neurodegenerative disorders; non-invasive brain stimulation; Alzheimer disease's; Parkinson's disease; autism; demyelinating disease; neuropsychology; ataxy

1. Introduction

Neuroimmunology has arisen as an innovative and integrative field in the shared edges of neurosciences, immunology, as well as related areas. The current research was carried out in order to get insight into the current development of neuroimmunology, in terms of scientists, events and literature shaping and reflecting its main achievements: the delineation and evolution of a central paradigm. A first topic addressed was to review some events on the history of neuroimmunology. This talk fixed the roots of neuroimmunology in the 19th century, with the descriptions of some of the model diseases (e.g., multiple sclerosis, neuromyelitisoptica, Guillain-Barré syndrome), the diagnostic examination of cerebrospinal fluid (CSF) by Quincke or the development of neuroimmunotherapy with the vaccine for rabies by Pasteur, and it showed that the field began to become delineated as an independent research area in the 1940s, after Medawar's statements on the peculiarities of the brain in antigenic tolerance and the findings by Waksman and Adams on experimental allergic (later coined as autoimmune) encephalomyelitis. Reviewed aspects of neuroimmunology were shown to have evolved from neurological infections and immunopathological conditions to the conformation of a supracontrol system integrating neural and immune mechanisms toward a better adaptation of the organism in an ever-changing environment, and this allowed for a definition of the purpose of the work and its significance in this field [1,2].

This meeting was the fourth edition of the International Workshop of Neuroimmunology and the first of the Cuban Network of Neuroimmunology "CUBANNI", going from the 10 to 14 June 2017. The first session chaired by Professor Von Wee Yong and Professor Maria Robinson opened with an atypical but very interesting talk from Prof. Orlando Serrano from Tunas University, Cuba. This talk offered an overview on the history of neuroimmunology. After that, some very experienced speakers in this field showed some advances in this field and discussed current topics relative to how the immune reaction occurs in the context of the neurological disorders, as well as discussing, in some cases, ways to restore functions or to modulate aberrant mechanisms to get beneficial effects that are disease related. Several diseases were revised, such as multiple sclerosis (MS), Alzheimer's disease, epilepsy, strock, autism, Parkinson disease, as well as others. Interesting lectures served as guides to offer a better understanding of the molecular processes underlying neuroimmune pathology in neurodegenerative, neurodevelopmental, and neurovascular diseases, as well as to understand some vulnerabilities and comorbidities occurring in these disorders. Relevant to this topic was the lecture of Prof. Jack Antel on a new approach in microRNA in MS vulnerabilities [1].

The papers that referred to new therapeutic approaches for interventions in neuroimmunology were presented by experts: Prof. Diogo O Souza from UFRGS, Brazil; Prof. Von Wee Yong from the University of Calgary, Canada, Prof. Luis Velazquez from the Center for the Research and Rehabilitation of Hereditary Ataxias (CIRAH), Holguín, Cuba, Prof. Lázaro Gómez from the International Center for Neurological Restoration (CIREN), Havana, Cuba, and Prof. Miriam Marañón Cardonne from the Applied Electromagnetism Center, Santiago de Cuba, Cuba. An experiment in the use of protector molecules in neurological diseases was also presented by Prof. Yanier Núñez-Figueredo from the Center for Drugs Research and Development (CIDEM), Havana City, Cuba.

2. Lecture Highlights

Manfred Schedlowski, from the Institute of Medical Psychology and Behavioral Immunobiology, University Clinic Essen, Germany, delivered his lecture entitled "Learned immune responses—mechanisms and potential clinical application".

This speaker introduced experimental data in rodent and humans demonstrating that peripheral humoral and cellular immune functions can be modulated by associative learning processes. In a taste-immune learning paradigm in rats employing the immunosuppressive drug cyclosporine A as an unconditioned stimulus, the behaviorally conditioned suppressive effects on IL-2 and IFN-γ mRNA expression, cytokine production and T cell activity were shown to be centrally mediated via the insular cortex and amygdala, and peripherally via the sympathetic innervation of lymphatic organs and the beta-adrenoreceptor-dependent inhibition of the calcineurin activity. This learned immunosuppression is of biological relevance, since behavioral conditioning significantly attenuated allergic responses and symptom severity in chronic inflammatory autoimmune diseases. Moreover, behaviorally conditioned immunosuppression has also been demonstrated in healthy humans and in diseases. The presented data provided proof of the principle that learning strategies can be implemented in a immunopharmacological regimen, the aim being to reduce the amount of drugs, thereby minimizing unwanted toxic drug side effects while maximizing the therapeutic benefit for patients [3,4].

Professor Wekerle from the Max Planx Institute, Germany entitled his lecture: "Nature and Nurture-Gut Microbiota and Multiple Sclerosis". This very experienced investigator commented on his work group's extensive experience, spanning multiple years, in using a spontaneous variant of relapsing-remitting experimental autoimmune encephalomyelitis (EAE) to study the triggering events of brain autoimmunity. Single transgenic JL/J mice expressing a myelin oligodendrocyte glycoprotein (MOG) showed autoreactive T cell receptor in about 70% of their CD4+ T cell repertoire. He showed how the disease risk is modulated by the modification of the microbiota, for instance by antibiotic treatments or by dietary interventions, as well as showing how they currently explore the effect of microbial samples from people with MS on initiation of EAE in the RR-mouse model. Prof. Wekerle's pioneering work in this field also exposed the previous results observed in a rodent model with spontaneous EAE, establishing the relevance of gut flora in disease-triggering demyelination [5–8]: gut-associated lymphoid tissue, gut microbes and the susceptibility to experimental autoimmune encephalomyelitis [9].

The human gut commensal microbiota forms a complex population of microorganisms that survive by maintaining a symbiotic relationship with the host. Amongst the metabolic benefits, it plays an important role in the formation of an adaptive immune system and in the maintenance of its homeostasis, and it is crucial in the induction, training, and function of the host immune system. An optimal immune system-microbiota alliance allows the induction of protective responses to pathogens and the maintenance of regulatory pathways to the tolerance to innocuous antigens. Nevertheless, a mutual impact of the gastrointestinal tract (GIT) and central nervous system (CNS) functions was recognized since the mid-twentieth century, and it has been accepted that the so-called gut-brain axis provides a two-way homeostatic communication, through immunological, hormonal and neuronal signals [10,11].

The GIT might represent a vulnerable area through which pathogens influence all aspects of physiology, even inducing CNS neuro-inflammation. A dysfunction of this axis has been associated with the pathogenesis of some diseases outside the GIT that have shown an increase in incidence over the last decades. Some papers had looked at how the commensal gut microbiota influences a systemic immune response in some neurological disorders, highlighting its impact cognition in multiple sclerosis, Guillain-Barrè syndrome, neurodevelopmental and behavioral disorders and Alzheimer's disease [12]. Nevertheless there no full comprehension on the implication of the potential microbiota-gut-brain dialogue in neurodegenerative diseases that might provide an insight on the pathogenesis and therapeutic strategies of these disorders.

The discussion of this talk focused on the fact that genes identified as having a susceptibility to MS do not satisfy the risk that argued for MS development. It was emphasized that many environmental risk factors to MS are described [13,14], but a growing number of studies demonstrate that gut microbes could directly play a role in the onset or progression of MS. A study led by researchers at the Max Planck Institute in Germany and co-authored by Cekanaviciute and Baranzini was referred to, in which it was found that microbiome transplants from MS patients could exacerbate symptoms in mice with a genetic model of the disease and which identified that specific species of bacteria more common in people with MS—*Akkermansia muciniphila* and *Acinetobacter calcoaceticus*—triggered the cells to become pro-inflammatory responses in human peripheral blood mononuclear cells and in monocolonized mice, while a species of bacteria found at lever that were lower than usual in MS patients—*Parabacteroides distasonis*—triggered immune-regulatory responses, stimulating anti-inflammatory IL-10-expressing human CD4$^+$CD25$^+$ T cells and IL-10$^+$FoxP3$^+$ Tregs in mice [13,14].

Finally, the experiment with microbiota transplants from MS patients to germ-free mice resulted in more severe symptoms of experimental autoimmune encephalomyelitis and reduced proportions of IL-10+ Tregs compared with mice that were "humanized" with microbiota from healthy controls. This study identified specific human gut bacteria that regulate adaptive autoimmune responses, suggesting a therapeutic targeting of the microbiota as a treatment for MS at time that provided evidence that MS-derived microbiota contain factors that precipitate an MS-like autoimmune disease in a transgenic mouse model.

Microbial colonization of the infant gastrointestinal tract (GIT) begins at birth, is shaped by the maternal microbiota, and is profoundly altered by antibiotictreatment [15,16]. Antibiotictreatment of mothers during pregnancy influences the colonization of the GIT microbiota of their infants [15]. The role of the GIT microbiota in regulating the adaptive immune function against systemic viral infections during infancy remains undefined. Nevertheless, an undisturbed colonization and progression of the GIT microbiota during infancy are necessary to promote robust adaptive immune responses. The topic of symbiotic and antibiotic interactions between gut commensal microbiota and the host immune system was discussed, regarding the integral elements of commensal microbiota that stimulate responses of different parts of the immune system and lead to health or disease, as well as on the conditions and factors that contribute to gut commensal microbiota's transformation from a symbiotic to an antibiotic relationship with humans [16]. The authors suggested to continue research efforts to better understand the host-microbiota interactions in physiological and disease settings, which might lead to the development of rational-based treatments.

For decades, the brain was considered an autonomous tissue that performs best without any assistance from the immune system. It is now widely accepted, largely through our work, that circulating monocytes and CD4+ T cells are needed for supporting brain reparation and functional plasticity. It was demonstrated that leukocytes could gain access to the brain's territory through a unique interface, the epithelial layer that forms the blood-CSF-barrier, the choroid plexus (CP) [5,8]. Michal Schwartz and Aleksandra Deczkowska, from The Weizmann Institute of Science, Rehovot, Israel, exposed these considerations in their lecture entitled "Reversing age-related dementia and Alzheimer's disease by restoring the brain-immune axis". Evidences from a Robust RNAseq analysis demonstrated that, in aging, the function of this interface is suppressed due to a systemic loss of interferon-gamma (IFN-γ), and due to a local elevation of type I interferon (type 1-IFN), which not only interferes with the activity of the CP but also negatively affects microglial activity. The intracerebral administration of anti-IFN-β receptor antibodies restored CP activity, reversed the microglial cognitive-impairing phenotype, and improved the overall cognitive ability in aged mice [17–20].

In an animal model of Alzheimer's disease, the CP was found to be dysfunctional, mainly due to low IFN-γ availability, which results from systemic immune suppression/exhaustion. Transiently reducing the systemic immunosuppression via the depletion of Foxp3+ regulatory T cells, or via the blocking of the inhibitory PD-1/PD-L1 immune checkpoint pathways, led to an increase of the activation of the CP to express trafficking molecules, which in turn led to the recruitment of immune regulatory cells to sites of brain pathology. Treatments with anti-PD-1 antibodies were found to be effective in several mouse models of AD in reversing cognitive loss, the removal of plaques, and in restoring brain homeostasis as determined by the inflammatory molecular profile. Such an approach is not meant to be against any single disease-escalating factor in AD, but rather empowers the individual's immune system to drive the process of repair. By directly targeting the immune system, rather than a single disease risk factor in the brain, this approach provides a comprehensive therapy that addresses numerous factors that become detrimental in the AD brain [21].

The role of the S100B protein, in neuroinflammation and Alzheimer's disease, was presented by Prof. Carlos-Alberto Gonçalves from the Federal University of Rio Grande do Sul, Puerto Alegre, Brazil. Alzheimer's disease (AD) is characterized by two main neuropathological brain patterns: the senile plaques formed by deposits of the β-amyloid protein (Aβ) at cholinergic terminals and the neurofibrillary tangles produced by tau protein hyper-phosphorilation. This evidence points to the fact that AD is not restricted to neurons; that neuroinflammation precedes AD pathogenesis; and that glial activation occurs early on in AD, even before cognitive deficit and amyloid deposition. Like microglia, astrocytes recognize danger-associated and pathogen-associated molecular patterns and release inflammatory mediators, as well as supporting and modulating neuronal activity. S100B is a calcium-binding protein mainly expressed in astrocytes in brain tissue, and experimental evidence indicates its involvement in neuroinflammation and degenerative disorders, including AD. This protein has many putative intracellular targets; it is secreted and has autocrine and paracrine effects on glia, neurons and microglia, while showing a toxic effect at a high concentration [22]. Some proposed intracellular targets of S100B include proteins of the cytoskeleton (e.g., GFAP), modulators of the cell cycle (e.g., p53 and Ndr kinase), the protein kinase C and phosphatase 2B (calcineurin). The extracellular effect of S100B, observed in neural cultures, depends on its concentration, since it is neurotrophic at pico and nanomolar levels and apoptotic at micromolar levels. The activation of RAGE and signaling pathways such as ERK, NF-κB and NFAT, following the effects of S100B observed in culture were discussed by Prof. Gonçalves, along with its role in the inflammatory response and its putative use as a biomarker in the three stages of AD [23–27].

The results of a novel study on the participation of neuroinflammatory processes in neurological pathologies such as epilepsy were shown. In this field, Dr. Lourdes Lorigados and collaborators presented clinical evidence of the involvement of inflammatory processes both centrally and peripherally in drug-resistant epilepsy with their work entitled: Two year follow-up of IL-1, IL-6 in temporal lobe drug-resistant epilepsy patient treated surgically [28,29].

Autism spectrum disorders are neurodevelopmental disorders which involve moderately to severely disrupted functioning in regard to social skills and socialization, expressive and receptive communication, and repetitive or stereotyped behaviors and interests [30]. "How Neuropsychology informs Neuroimmunology—Understanding the Cognitive and Behavioral Vulnerabilities in Autism Spectrum Disorders", was the lecture delivered by Prof. Scott Hunter of Chicago University, USA. This presentation addressed the current understanding of the neuropsychology of autism spectrum disorders (ASD), in association with ongoing research on neuroimmunological factors that contribute to higher risks of neurodevelopmental disorders. The spectrum of challenges experienced by children with ASD and the multiple levels of influence that occur neurodevelopmentally, both prenatally and postnatally, were reviewed [31]. A proposal for an integrated collaboration between researchers and clinicians was also offered, to guide considerations regarding etiology and treatment in this disorder.

"From clinical to molecular in autism: A pending task to the science", delivered by Prof. Maria Robinson Agramonte from the International Center for Neurological Restoration, Havana, Cuba, was based on a study involving clinical and molecular analysis. It showed how necessary it became to perform multidisciplinary studies in autism in order to advance the understanding of the disease pathology toward better clinical management and toward the implementation of novel therapies in development. Referring to the contribution of inflammatory markers to vulnerabilities in autism, this paper showed the potentialities of EEG findings in aiding information on clinical decision making for children with autism spectrum disorders [32].

Dalina Laffita from Florida Atlantic University, delivered a talk entitled "Suppressing Igf-1 signaling pathway rescues neuronal overgrowth of Pten+/- mice". Macrocephaly is associated with ASD. The alteration of mTOR signaling pathways has been shown to be involved in 14% of ASD individuals [33]. The research supports divergent trajectories of brain growth in the regressive vs. non-regressive phenotype of ASD; however, the degree to which the cellular mechanisms underlying these profiles differ is not clear [34]. PTEN is a gene associated with ASD, the macrocephaly is observed in the heterozygous mutation in PTEN (PTEN+/-), which leads to a negative regulation in PI3K/Akt/mTOR signaling pathways; it is a great model for studying ASD, as these mice show social behavioral deficits that are observed in humans. During this study, there was evidence that Pten is a regulator of β-catenin. An increase was seen in the β-catenin signal, but the manner in how both of these are associated is not well understood. Measurements of the soma cell size of the fifth layer of the cerebral cortex in a Pten+/- mouse were taken. There was a decrease in the sizes of the neurons; thus, the ultimate goal was to recover the original size after birth, by upregulating the PI3K/Akt/mTOR signaling pathway. The inhibition of the insulin-like growth factor receptor 1 (IGF1R), which is found upstream of this pathway, led to a significant decrease in soma cell size in Pten+/-/Igf-1r+/- mice. The authors considered this a very promising result because it opens the door for a new pharmacological agent that could recover the damage done by this mutation. The pathway being targeted is a cellular pathway, so this brings up the challenge that the pharmacological agent must be able to cross the blood brain barrier and be able to only influence this cellular pathway within the brain to prevent side effects.

Georg Auburger from Goethe University Hospital Frankfurt, Germany, delivered a lecture on Ataxin-2 as a translation factor with a specificity for mitochondrial precursors. Prof. Auburger introduced his lecture by referencing the Ataxin-2 (ATXN2) gene family characterization, saying that, via yeast, it extends from man to plants and that it appears to be so important for all forms of eukaryotic life that several gene copies exist in most vertebrates, as well as in plants and fungi. Mammalian Ataxin-2 is transcriptionally induced during starvation; he said and continued saying that in periods of nutrient depletion, oxidative stress or infection, its subcellular localization changes from the ribosomal translation apparatus to stress granules, where RNA quality control and RNA repair occur [35].

Ataxin-2 is a key element of the stress granules component. Over the last two years, it was shown in yeast, worms and mammals that Ataxin-2 represses global mRNA translation via diminished mTORC1 signaling, a pathway where several drugs are already available. Importantly however, the abundance of selected factors requires Ataxin-2 expression. One example is the PAS domain containing the factor period. The PAS domains are signal modules that are widely distributed in proteins across all kingdoms of life. They are common in photoreceptors and transcriptional regulators of eukaryotic circadian clocks and mainly possess protein-protein interaction and light-sensing functions [36], which triggers rest and the decline of metabolic activity during nighttime. Another example is the mitochondrial matrix enzyme Iso-Valeryl-Dehydrogenase, which is responsible for leucine degradation and thus reduces mTORC1 activity. Further examples are branched chain amino acid enzymes in the mitochondrial matrix, which are responsible for fatty acid degradation. Additionally, mitochondrial matrix enzymes in the tricarboxylic acid (TCA) cycle of glycolysis are enhanced by Ataxin-2 expression. When Ataxin-2 is not absent, fat accumulates subcutaneously and in liver lipid droplets, parallel to glycogen granules. Thus, the recruitment of stored alternative fuels in times of glucose depletion or of excessive bioenergetic demands is dependent on Ataxin-2. Given that cellular lipid uptake via the endocytosis of trophic receptors is impaired by Ataxin-2, cells are more dependent on their own nutrient reserves, and autophagy would probably increase. Interestingly, Ataxin-2 expression also enhances the abundance of PINK1, an autophagy-regulator that is responsible for the quality control and repair of mitochondria. A conclusion of this lecture was that these cellular roles seem essential for neurodegenerative diseases, given that ATXN2 depletion is surprisingly effective in protecting motor neurons and cerebellar neurons in ALS and in Spinocerebellar Ataxias (SCAs), as demonstrated in mouse models and flies.

Spinocerebellar Ataxias (SCA) are neurodegenerative disorders characterized by the degeneration of the cerebellum and its afferent and efferent connections, brainstem, spinal cord and peripheral nerves. In the past 20 years, clinical trials have been carried out in SCAs to slow down or stop disease progression and disability. However, the main findings were often negative due to the fact that the included patients had a severe degeneration of the nervous system. The best moment to start with clinical trials is before ataxia onset, where the neurodegeneration is beginning. Recently, two phases before ataxia onset were defined: the asymptomatic and the preclinical stages in SCAs. The asymptomatic phase is characterized by an absence of neurological abnormalities. However, in the preclinical phase, there are several clinical and paraclinical complaints in SCAs. "Spinocerebellar Ataxias: Early diagnosis and ethical dilemmas during preclinical trials", was presented by Prof. Luis Velázquez-Pérez from the Center for Research and Rehabilitation of Hereditary Ataxias, Holguin, Cuba. In this talk, he commented on how the first studies of the preclinical stage were carried out in a Cuban SCA2 population. The dorsal lemniscal system, saccades movements, visual motor performance, REM sleep, and autonomic nervous system were evaluated. 35% of the preclinical subjects have cerebellar alterations such as abnormal tandem gait, mild incoordination deficit and nistagmus. The non-cerebellar manifestations were muscle cramps, sensory symptoms and sleep disorders. Clinical examination showed hyperreflexia, saccade slowing and cognitive dysfunction. The early clinical trials raise ethical dilemmas which, if not effectively addressed, may harm subjects. The high prevalence of SCA2 families in Cuba has offered great opportunities to carry intervention strategies in subjects in a preclinical state. SCA2 preclinical carriers revealed a high acceptation rate (97.5%) to be included in neurorehabilitation programs and/or clinical trials. Thus, Cuban SCA2 preclinical carriers have undergone early therapeutical approaches consisting in a rehabilitation program and an open-label clinical trial with vitamins, which improved both the subtle cerebellar manifestations and muscle cramps. Both treatments were safe in all subjects, minimizing the ethical dilemma associated with the drug side effects. Regarding the ethical concerns related to the placement of a preclinical carrier on a placebo group, all interviewed subjects that were referred agreed with these designs due to the possibility of receiving concomitant therapeutical options during the study and receiving the true drug after completion of the study, if it was effective and safe. In addition, the ethical dilemma of disease stigmatization were not observed in the treated SCA2 preclinical carriers, which could be explained by the long psychotherapeutical follow-up treatment that they received. In summary, the ethical concerns raised from the early interventions in prodromal SCA2 have led to new challenges for physicians, genetic counselors and researchers, which must be addressed via further investigations [37–39].

The cognitive impairment reported in 25% of patients is a common feature of Spinocerebellar Ataxia type 2 (SCA2), but has not been systematically studied, with no surrogate biomarkers having been described. In this context, Roberto Rodríguez-Labrada, also coming from the Ataxia Center in Cuba, exposed the results of a study guide to characterize new biomarkers of cognitive decline in SCA2 through clinical and electrophysiological assessments. The lecture, entitled "New biomarkers for Spinocerebellar Ataxia type 2: insight into physiopathology of cognitive decline", included the INECO frontal screening test and an antisaccadic eye movement paradigm in 40 SCA2 patients, 37 preclinical mutation carriers and 40 healthy controls. The results evidenced a significant decrease of mean INECO scores in SCA2 patients and preclinical carriers, when compared to controls. This score was inversely correlated to expanded CAG repeats and the SARA score in the SCA2 patients, and was directly correlated to the time to ataxia onset in the preclinical carriers. Regarding the antisaccadic assessment, both disease related groups exhibited increased antisaccadic error rates and latencies, but the percentage of antisaccadic errors correction was only reduced in the patients. SCA2 patients carrying larger CAG repeats showed poorer antisaccadic performances, while preclinical carriers with shorter times to ataxia onset exhibited larger antisaccadic latencies [40–42]. A conclusion of this study reveals the usefulness of INECO tests and antisaccadic movements as sensitive biomarkers of executive dysfunctions in SCA2 since the early stages of the disease, as well as offering new evidence on the role of the cerebellum in cognitive functions from a human model disease.

"Metabolic mismatches and spinocerebellar Ataxia type 2 clinical phenotype": Almaguer-Gotay Dennis from the Center for Research and Rehabilitation of Hereditary Ataxias (CIRAH) showed the results of his group, related to the characterization of the SCA2 metabolism in presymptomatic patients. They concluded that SCA2 have some metabolic mismatches beginning from the presymtomatic stages, which have a significant correlation with disease severity and progression markers [43,44]. Almaguer Mederos LE, from the same center (CIRAH), claimed that *MTHFR* C677T polymorphism is associated with disease progression in SCA 2. These results point to a possible implication of the folate metabolism in the pathophysiology of SCA2, having a role in disease progression but not in disease onset [43].

Prof. Ivón Pedroso Ibáñez delivered her lecture entitled "Erythropoietin induce neuroprotection in patients with Parkinson's disease". Dr. Pedroso discussed the results of a multicenter study at CIREN (International Center for Neurological Restauration), CIM (Center for Molecular Immunology) and CEMPALA (Center for Attention and Development of Animal for laboratory, Cuba), which showed that treatment strategies in Parkinson's disease (PD) have improved patients' quality of life without stopping their progression, and that different alternatives to modify the natural course of the disease have demonstrated neuroprotective properties of erythropoietin; particularly, Cuban recombinant form (EPOrh) and recombinant human erythropoietin with low sialic acid (NeuroEPO) have showed good results. This talk showed the results of two clinical trials for neuroprotection purposes performed by this group using patients with PD in the IV stages of the Hoenh and Yarh scale, who received an intranasal formulation of NeuroEPO. All patients were evaluated with a battery of neuropsychological scales Data showed that both molecules are tolerated by patients with Parkinson's disease, with only mild adverse effects and a positive response to the cognitive functions. Authors considered that more studies must be carry out to underline the real relevance of this new biotechnological product. Additionally, these results contribute to the necessary screening of these types of compounds, promising protective properties to neurodegenerative diseases [45,46].

In addition, the results of an intra-nasal administration of non-hematopoietic erythropoietin promoting a clinical benefit without toxicity in spinocerebellar ataxia type 2 patients were discussed from a double-blind experiment; a placebo-controlled phase 2 clinical trial presented by Prof. Luis Velázquez and colleagues from the Center for the Research and Rehabilitation of Hereditary Ataxias, Holguín, Cuba. This study demonstrated that NeuroEpo is safe, feasible, and it suggested that it might represent a new therapeutic strategy for SCA and other neurodegenerative diseases [46].

Currently the biotechnological industry offers novel potential molecules with neuroprotective properties. An area of the meeting was reserved for the presentations of Prof. Diogo O Souza, from UFRGS, Brasil, who showed the results of his group on the neuroprotective effect of guanosine in the experimental models of brain diseases. Glutamate is the main excitatory neurotransmitter in mammalian CNS, essential for most brain activities. However, the hyperactive activation of the glutamatergic system may be potentially neurotoxic, involving the pathogenesis of various acute and chronic brain injuries. Increased glutamate causes the sustained entry of calcium and its excessive accumulation inside the cell, firing various intracellular processes that eventually lead to cell death. This increase of calcium generates the formation of free radicals that promote lipid peroxidation at the level of the membranes and the synthesis of nitric oxide, which acts as feedback and powers the excitotoxic effect by increasing the release of glutamate.

ALS is an age-related neurodegenerative disorder that is believed to have complex genetic and environmental influences in pathogenesis, but the etiologies are unidentified for most patients. Until the major causes are better defined, drug development is directed at downstream pathophysiological mechanisms, which are themselves incompletely understood. For nearly 30 years, glutamate-induced excitotoxicity has lain at the core of theories behind the spiraling events, including mitochondrial dysfunction, oxidative stress, and protein aggregation, that lead to neurodegenerative cell death in ALS [47]. The main process responsible for maintaining the extracellular glutamate levels below the toxic concentration is the glutamate uptake exerted by glutamate transporters located in neural cell membranes, mainly in astrocytes. This group showed strong evidence that the guanine-based purinergic system is effectively neuroprotective against glutamate toxicity, in acute and chronic animal models from both in vitro and in vivo studies. These results indicate that the neuroprotective guanine-based purine is the nucleoside guanosine (Guo). In vivo studies showed that Guo (i.c.v., i.p. or orally administered) protects against seizures (induced by quinolinic acid), brain ischemia and hepatic encephalopathy. The group searched for mechanisms implicated in this neuroprotection and demonstrated that this compound stimulates the astrocytic glutamate uptake in astrocyte cultures, additionally exerting neuroprotective effects that avoid the decrease in glutamate uptake [23,24,48,49].

Other molecules with neuroprotective effects were exposed by Wong-Guerra Maylin from the Center for the Development and Research of Medications, Cuba, with her paper entitled "Neuroprotective effects of JM-20 on aluminium-chloride-induced learning impairments, mitochondrial dysfunction and neuronal death in adult rats". JM-20 is a hybrid benzodiazepine dihydropyridine molecule, with a neuroprotective effect demonstrated in brain disorders involving excitotoxicity, oxidative damage, inflammatory response and mitochondrial dysfunction. This paper showed that the treatment with JM-20 prevented the increase of the expression of pro-apoptotic signaling proteins, protecting neuronal loss and presumably cognitive impairment, and in this line of findings the authors suggested that JM-20 may prevent memory impairment via mitochondrial protection and the inhibition of the intrinsic apoptosis pathway, also suggesting a therapeutic benefit to neurodegenerative dementia like Alzheimer's disease. The results on the regional effects of transient cerebral ischemia on astrocytes reactivity and the PI3K/Akt survival pathway, and the role of JM-20, were shown by Ramírez-Sánchez Jeney [50,51].

A Cuban experiment on a novel tool of intervention, "Transcraneal Magnetic Stimulation", was presented by Prof. Lázaro Gómez's lecture, "Effects of Non-Invasive Brain Stimulation in children with Autism Spectrum Disorder: a short term outcome study".

From his approach, the core symptoms of ASD may improve with pharmacological and non-pharmacological interventions; but deficits in social interaction and language remain throughout the lifespan. Following this criteria, the study showed how non-invasive brain stimulation (NIBS) could be a valuable adjuvant therapy for ASD, from the evaluation of tolerability, safety and the clinical effect of NIBS in children with ASD. The Autism Diagnostic Interview Revisited (ADI-R), Autism Treatment Evaluation Checklist (ATEC) and the Autism Behaviour Checklist (ABC) were used to assess the intervention effects, comparing the scores before and one week, one month, three months and six months after completing all the sessions. The resting state EEG was based in the analysis of functional connectivity, which showed clear changes after treatment. A conclusion of this lecture was that both methods, tDCS and rTMS, were tolerable and safe for the patients included in the study, and that a remarkable clinical improvement in their autistic symptoms was found after treatment [52].

3. Conclusions

The prevalence of inflammatory processes became a major component in neurological disease, at time when a main focus in neuroscience is to look for both a major understanding of neuroimmune pathology in the context of these diseases, as well as more effective and earlier ways of intervention. Following this viewpoint, this meeting consistently highlighted the discussion of molecular mechanisms and therapeutic approaches related to a more effective neuromodulation in neurological diseases, with particular relevance to neurodevelopment and neurodegenerative diseases. Results on The TMS and NeuroEpo offered a good perspective on neuroprotectors and modulators of neuroplasticity, since the C-PC and PCB were shown to be useful and promising therapeutic options for MS, based in the remyelinating effect observed in EAE models. The microbiome was treated as a third player in MS pathogenesis. Single-target approaches are insufficient for these multifactorial diseases, and molecules are screened for different relevant targets and effects. As we observed before, natural products continue to be an important source of drugs with beneficial effects for NDs; the roles of oxidative stress and mitochondrial performance are strongly linked to AD pathogenesis; in silico research is the leading approach in drug discovery and it is being widely used, mainly for AD. The different methodologies and lines of research discussed in the symposium suggest new possible therapeutic options and provide further insights on NDs mechanisms, from basic to translational research, drug discovery (mainly from natural sources), and the use of high-technology methods. The hallmarks of the symposium were all focused on NDs and neurodevelopment disorders. The congress provided a framework for translational research to advance toward new promising neural disease-modifying therapies.

Author Contributions: M.d.l.A.R.-A. was President of the Organizing Committee of International Conference "CUBANNI 2017" and elaborated the conference report. L.L.P. was President of the Scientific Committee and participated writing and reviewing the conference report. O.R.S.-B. participated reviewing the conference report.

Funding: This research received no external funding.

Acknowledgments: The authors want to acknowledge speakers as the main contributors of this paper and the editorial committee for this special issue.

Conflicts of Interest: The authors declare no conflicts of interest.

References

1. Ader, R.; Kelley, K.W. A global view of twenty years of brain, behavior, and immunity. *Brain Behav. Immun.* **2007**, *21*, 20–22. [CrossRef] [PubMed]
2. Wekerle, H. T-cell autoimmunity in the central nervous system. *Intervirology* **1993**, *35*, 95–100. [CrossRef] [PubMed]

3. Schedlowski, M.; Pacheco-Lopez, G. The learned immune response: Pavlov and beyond. *Brain Behav. Immun.* **2010**, *24*, 176–185. [CrossRef] [PubMed]
4. Albring, A.; Wendt, L.; Benson, S.; Nissen, S.; Yavuz, Z.; Engler, H.; Witzke, O.; Schedlowski, M. Preserving learned immunosuppressive placebo response: Perspectives for clinical application. *Clin. Pharmacol. Ther.* **2014**, *96*, 247–255. [CrossRef] [PubMed]
5. Hohlfeld, R.; Dornmair, K.; Meinl, E.; Wekerle, H. The search for the target antigens of multiple sclerosis, part 1: Autoreactive CD4$^+$ T lymphocytes as pathogenic effectors and therapeutic targets. *Lancet Neurol.* **2016**, *15*, 198–209. [CrossRef]
6. Hohlfeld, R.; Dornmair, K.; Meinl, E.; Wekerle, H. The search for the target antigens of multiple sclerosis, part 2: CD8$^+$ T cells, B cells, and antibodies in the focus of reverse-translational research. *Lancet Neurol.* **2016**, *15*, 317–331. [CrossRef]
7. Hohlfeld, R.; Wekerle, H. Multiple sclerosis and microbiota. From genome to metagenome? *Nervenarzt* **2015**, *86*, 925–933. [CrossRef] [PubMed]
8. Wekerle, H. Brain autoimmunity and intestinal microbiota: 100 trillion game changers. *Trends Immunol.* **2017**, *38*, 483–497. [CrossRef] [PubMed]
9. Wekerle, H. The gut-brain connection: Triggering of brain autoimmune disease by commensal gut bacteria. *Rheumatology* **2016**, *55*, ii68–ii75. [CrossRef] [PubMed]
10. Belkaid, Y.; Artis, D. Immunity at the barriers. *Eur. J. Immunol.* **2013**, *43*, 3096–3097. [CrossRef] [PubMed]
11. Belkaid, Y.; Hand, T.W. Role of the microbiota in immunity and inflammation. *Cell* **2014**, *157*, 121–141. [CrossRef] [PubMed]
12. Catanzaro, R.; Anzalone, M.; Calabrese, F.; Milazzo, M.; Capuana, M.; Italia, A.; Occhipinti, S.; Marotta, F. The gut microbiota and its correlations with the central nervous system disorders. *Panminerva Med.* **2015**, *57*, 127–143. [PubMed]
13. Cekanaviciute, E.; Yoo, B.B.; Runia, T.F.; Debelius, J.W.; Singh, S.; Nelson, C.A.; Kanner, R.; Bencosme, Y.; Lee, Y.K.; Hauser, S.L.; et al. Gut bacteria from multiple sclerosis patients modulate human T cells and exacerbate symptoms in mouse models. *Proc. Natl. Acad. Sci. USA* **2017**, *114*, 10713–10718. [CrossRef] [PubMed]
14. Berer, K.; Gerdes, L.A.; Cekanaviciute, E.; Jia, X.; Xiao, L.; Xia, Z.; Liu, C.; Klotz, L.; Stauffer, U.; Baranzini, S.E.; et al. Gut microbiota from multiple sclerosis patients enables spontaneous autoimmune encephalomyelitis in mice. *Proc. Natl. Acad. Sci. USA* **2017**, *114*, 10719–10724. [CrossRef] [PubMed]
15. Gonzalez-Perez, G.; Hicks, A.L.; Tekieli, T.M.; Radens, C.M.; Williams, B.L.; Lamouse-Smith, E.S. Maternal antibiotic treatment impacts development of the neonatal intestinal microbiome and antiviral immunity. *J. Immunol.* **2016**, *196*, 3768–3779. [CrossRef] [PubMed]
16. Malys, M.K.; Campbell, L.; Malys, N. Symbiotic and antibiotic interactions between gut commensal microbiota and host immune system. *Medicina* **2015**, *51*, 69–75. [CrossRef] [PubMed]
17. Deczkowska, A.; Baruch, K.; Schwartz, M. Type I/II interferon balance in the regulation of brain physiology and pathology. *Trends Immunol.* **2016**, *37*, 181–192. [CrossRef] [PubMed]
18. Deczkowska, A.; Matcovitch-Natan, O.; Tsitsou-Kampeli, A.; Ben-Hamo, S.; Dvir-Szternfeld, R.; Spinrad, A.; Singer, O.; David, E.; Winter, D.R.; Smith, L.K.; et al. Mef2c restrains microglial inflammatory response and is lost in brain ageing in an IFN-I-dependent manner. *Nat. Commun.* **2017**, *8*, 717. [CrossRef] [PubMed]
19. Schwartz, M.; Deczkowska, A. Neurological disease as a failure of brain-immune crosstalk: The multiple faces of neuroinflammation. *Trends Immunol.* **2016**, *37*, 668–679. [CrossRef] [PubMed]
20. Baruch, K.; Ron-Harel, N.; Gal, H.; Deczkowska, A.; Shifrut, E.; Ndifon, W.; Mirlas-Neisberg, N.; Cardon, M.; Vaknin, I.; Cahalon, L.; et al. CNS-specific immunity at the choroid plexus shifts toward destructive TH2 inflammation in brain aging. *Proc. Natl. Acad. Sci. USA* **2013**, *110*, 2264–2269. [CrossRef] [PubMed]
21. Baruch, K.; Deczkowska, A.; Rosenzweig, N.; Tsitsou-Kampeli, A.; Sharif, A.M.; Matcovitch-Natan, O.; Kertser, A.; David, E.; Amit, I.; Schwartz, M. PD-1 immune checkpoint blockade reduces pathology and improves memory in mouse models of alzheimer's disease. *Nat. Med.* **2016**, *22*, 135–137. [CrossRef] [PubMed]
22. Michetti, F.; Corvino, V.; Geloso, M.C.; Lattanzi, W.; Bernardini, C.; Serpero, L.; Gazzolo, D. The S100B protein in biological fluids: More than a lifelong biomarker of brain distress. *J. Neurochem.* **2012**, *120*, 644–659. [CrossRef] [PubMed]

23. Bellaver, B.; Dos Santos, J.P.; Leffa, D.T.; Bobermin, L.D.; Roppa, P.H.A.; da Silva Torres, I.L.; Goncalves, C.A.; Souza, D.O.; Quincozes-Santos, A. Systemic inflammation as a driver of brain injury: The astrocyte as an emerging player. *Mol. Neurobiol.* **2017**, *55*, 2685–2695. [CrossRef] [PubMed]

24. Bellaver, B.; Souza, D.G.; Bobermin, L.D.; Goncalves, C.A.; Souza, D.O.; Quincozes-Santos, A. Guanosine inhibits LPS-induced pro-inflammatory response and oxidative stress in hippocampal astrocytes through the heme oxygenase-1 pathway. *Purinergic Signal.* **2015**, *11*, 571–580. [CrossRef] [PubMed]

25. De Souza, D.F.; Wartchow, K.; Hansen, F.; Lunardi, P.; Guerra, M.C.; Nardin, P.; Goncalves, C.A. Interleukin-6-induced S100B secretion is inhibited by haloperidol and risperidone. *Prog. Neuro-Psychopharmacol. Biol. Psychiatry* **2013**, *43*, 14–22. [CrossRef] [PubMed]

26. Guerra, M.C.; Tortorelli, L.S.; Galland, F.; Da Re, C.; Negri, E.; Engelke, D.S.; Rodrigues, L.; Leite, M.C.; Goncalves, C.A. Lipopolysaccharide modulates astrocytic S100B secretion: A study in cerebrospinal fluid and astrocyte cultures from rats. *J. Neuroinflamm.* **2011**, *8*, 128. [CrossRef] [PubMed]

27. Neves, J.D.; Aristimunha, D.; Vizuete, A.F.; Nicola, F.; Vanzella, C.; Petenuzzo, L.; Mestriner, R.G.; Sanches, E.F.; Goncalves, C.A.; Netto, C.A. Glial-associated changes in the cerebral cortex after collagenase-induced intracerebral hemorrhage in the rat striatum. *Brain Res. Bull.* **2017**, *134*, 55–62. [CrossRef] [PubMed]

28. Lorigados Pedre, L.; Morales Chacon, L.M.; Orozco Suarez, S.; Pavon Fuentes, N.; Estupinan Diaz, B.; Serrano Sanchez, T.; Garcia Maeso, I.; Rocha Arrieta, L. Inflammatory mediators in epilepsy. *Curr. Pharm. Des.* **2013**, *19*, 6766–6772. [CrossRef] [PubMed]

29. Lorigados Pedre, L.; Morales Chacon, L.M.; Pavon Fuentes, N.; Robinson Agramonte, M.L.A.; Serrano Sanchez, T.; Cruz-Xenes, R.M.; Diaz Hung, M.L.; Estupinan Diaz, B.; Baez Martin, M.M.; Orozco-Suarez, S. Follow-up of peripheral IL-1beta and IL-6 and relation with apoptotic death in drug-resistant temporal lobe epilepsy patients submitted to surgery. *Behav. Sci.* **2018**, *8*, 21. [CrossRef] [PubMed]

30. Lahiri, D.K.; Sokol, D.K.; Erickson, C.; Ray, B.; Ho, C.Y.; Maloney, B. Autism as early neurodevelopmental disorder: Evidence for an sappalpha-mediated anabolic pathway. *Front. Cell. Neurosci.* **2013**, *7*, 94. [CrossRef] [PubMed]

31. Sharma, S.; Woolfson, L.M.; Hunter, S.C. Maladaptive cognitive appraisals in children with high-functioning autism: Associations with fear, anxiety and theory of mind. *Autism Int. J. Res. Pract.* **2014**, *18*, 244–254. [CrossRef] [PubMed]

32. Inga Jacome, M.C.; Morales Chacon, L.M.; Vera Cuesta, H.; Maragoto Rizo, C.; Whilby Santiesteban, M.; Ramos Hernandez, L.; Noris Garcia, E.; Gonzalez Fraguela, M.E.; Fernandez Verdecia, C.I.; Vegas Hurtado, Y.; et al. Peripheral inflammatory markers contributing to comorbidities in autism. *Behav. Sci.* **2016**, *6*, 29. [CrossRef] [PubMed]

33. Nordahl, C.W.; Lange, N.; Li, D.D.; Barnett, L.A.; Lee, A.; Buonocore, M.H.; Simon, T.J.; Rogers, S.; Ozonoff, S.; Amaral, D.G. Brain enlargement is associated with regression in preschool-age boys with autism spectrum disorders. *Proc. Natl. Acad. Sci. USA* **2011**, *108*, 20195–20200. [CrossRef] [PubMed]

34. De Vries, P.J. Targeted treatments for cognitive and neurodevelopmental disorders in tuberous sclerosis complex. *Neurother. J. Am. Soc. Exp. NeuroTher.* **2010**, *7*, 275–282. [CrossRef] [PubMed]

35. Auburger, G.; Sen, N.E.; Meierhofer, D.; Basak, A.N.; Gitler, A.D. Efficient prevention of neurodegenerative diseases by depletion of starvation response factor ataxin-2. *Trends Neurosci.* **2017**, *40*, 507–516. [CrossRef] [PubMed]

36. Mei, Q.; Dvornyk, V. Evolution of pas domains and pas-containing genes in eukaryotes. *Chromosoma* **2014**, *123*, 385–405. [CrossRef] [PubMed]

37. Figueroa, K.P.; Coon, H.; Santos, N.; Velazquez, L.; Mederos, L.A.; Pulst, S.M. Genetic analysis of age at onset variation in spinocerebellar ataxia type 2. *Neurol. Genet.* **2017**, *3*, e155. [CrossRef] [PubMed]

38. Velazquez-Perez, L.; Rodriguez-Labrada, R.; Torres-Vega, R.; Montero, J.M.; Vazquez-Mojena, Y.; Auburger, G.; Ziemann, U. Central motor conduction time as prodromal biomarker in spinocerebellar ataxia type 2. *Mov. Disord. Off. J. Mov. Disord. Soc.* **2016**, *31*, 603–604. [CrossRef] [PubMed]

39. Velazquez-Perez, L.; Tunnerhoff, J.; Rodriguez-Labrada, R.; Torres-Vega, R.; Belardinelli, P.; Medrano-Montero, J.; Pena-Acosta, A.; Canales-Ochoa, N.; Vazquez-Mojena, Y.; Gonzalez-Zaldivar, Y.; et al. Corticomuscular coherence: A novel tool to assess the pyramidal tract dysfunction in spinocerebellar ataxia type 2. *Cerebellum* **2017**, *16*, 602–606. [CrossRef] [PubMed]

40. Rodriguez-Labrada, R.; Vazquez-Mojena, Y.; Canales-Ochoa, N.; Medrano-Montero, J.; Velazquez-Perez, L. Heritability of saccadic eye movements in spinocerebellar ataxia type 2: Insights into an endophenotype marker. *Cerebellum Ataxias* **2017**, *4*, 19. [CrossRef] [PubMed]

41. Rodriguez-Labrada, R.; Velazquez-Perez, L.; Aguilera-Rodriguez, R.; Seifried-Oberschmidt, C.; Pena-Acosta, A.; Canales-Ochoa, N.; Medrano-Montero, J.; Estupinan-Rodriguez, A.; Vazquez-Mojena, Y.; Gonzalez-Zaldivar, Y.; et al. Executive deficit in spinocerebellar ataxia type 2 is related to expanded cag repeats: Evidence from antisaccadic eye movements. *Brain Cogn.* **2014**, *91*, 28–34. [CrossRef] [PubMed]

42. Rodriguez-Labrada, R.; Velazquez-Perez, L.; Auburger, G.; Ziemann, U.; Canales-Ochoa, N.; Medrano-Montero, J.; Vazquez-Mojena, Y.; Gonzalez-Zaldivar, Y. Spinocerebellar ataxia type 2: Measures of saccade changes improve power for clinical trials. *Mov. Disord. Off. J. Mov. Disord. Soc.* **2016**, *31*, 570–578. [CrossRef] [PubMed]

43. Almaguer-Gotay, D.; Almaguer-Mederos, L.E.; Aguilera-Rodriguez, R.; Rodriguez-Labrada, R.; Cuello-Almarales, D.; Estupinan-Dominguez, A.; Velazquez-Perez, L.C.; Gonzalez-Zaldivar, Y.; Vazquez-Mojena, Y. Spinocerebellar ataxia type 2 is associated with the extracellular loss of superoxide dismutase but not catalase activity. *Front. Neurol.* **2017**, *8*, 276. [CrossRef] [PubMed]

44. Almaguer-Mederos, L.E.; Almaguer-Gotay, D.; Aguilera-Rodriguez, R.; Gonzalez-Zaldivar, Y.; Cuello-Almarales, D.; Laffita-Mesa, J.; Vazquez-Mojena, Y.; Zayas-Feria, P.; Rodriguez-Labrada, R.; Velazquez-Perez, L.; et al. Association of glutathione S-transferase omega polymorphism and spinocerebellar ataxia type 2. *J. Neurol. Sci.* **2017**, *372*, 324–328. [CrossRef] [PubMed]

45. Pedroso, I.; Bringas, M.L.; Aguiar, A.; Morales, L.; Alvarez, M.; Valdes, P.A.; Alvarez, L. Use of cuban recombinant human erythropoietin in parkinson's disease treatment. *MEDICC Rev.* **2012**, *14*, 11–17. [PubMed]

46. Santos-Morales, O.; Diaz-Machado, A.; Jimenez-Rodriguez, D.; Pomares-Iturralde, Y.; Festary-Casanovas, T.; Gonzalez-Delgado, C.A.; Perez-Rodriguez, S.; Alfonso-Munoz, E.; Viada-Gonzalez, C.; Piedra-Sierra, P.; et al. Nasal administration of the neuroprotective candidate neuroepo to healthy volunteers: A randomized, parallel, open-label safety study. *BMC Neurol.* **2017**, *17*, 129. [CrossRef] [PubMed]

47. Blasco, H.; Mavel, S.; Corcia, P.; Gordon, P.H. The glutamate hypothesis in als: Pathophysiology and drug development. *Curr. Med. Chem.* **2014**, *21*, 3551–3575. [CrossRef] [PubMed]

48. Hansel, G.; Ramos, D.B.; Delgado, C.A.; Souza, D.G.; Almeida, R.F.; Portela, L.V.; Quincozes-Santos, A.; Souza, D.O. The potential therapeutic effect of guanosine after cortical focal ischemia in rats. *PLoS ONE* **2014**, *9*, e90693. [CrossRef] [PubMed]

49. Quincozes-Santos, A.; Bobermin, L.D.; Tramontina, A.C.; Wartchow, K.M.; Tagliari, B.; Souza, D.O.; Wyse, A.T.; Goncalves, C.A. Oxidative stress mediated by NMDA, AMPA/KA channels in acute hippocampal slices: Neuroprotective effect of resveratrol. *Toxicol. In Vitro* **2014**, *28*, 544–551. [CrossRef] [PubMed]

50. Nunez-Figueredo, Y.; Pardo Andreu, G.L.; Oliveira Loureiro, S.; Ganzella, M.; Ramirez-Sanchez, J.; Ochoa-Rodriguez, E.; Verdecia-Reyes, Y.; Delgado-Hernandez, R.; Souza, D.O. The effects of JM-20 on the glutamatergic system in synaptic vesicles, synaptosomes and neural cells cultured from rat brain. *Neurochem. Int.* **2015**, *81*, 41–47. [CrossRef] [PubMed]

51. Ramirez-Sanchez, J.; Simoes Pires, E.N.; Nunez-Figueredo, Y.; Pardo-Andreu, G.L.; Fonseca-Fonseca, L.A.; Ruiz-Reyes, A.; Ochoa-Rodriguez, E.; Verdecia-Reyes, Y.; Delgado-Hernandez, R.; Souza, D.O.; et al. Neuroprotection by JM-20 against oxygen-glucose deprivation in rat hippocampal slices: Involvement of the AKT/GSK-3beta pathway. *Neurochem. Int.* **2015**, *90*, 215–223. [CrossRef] [PubMed]

52. Gomez, L.; Vidal, B.; Maragoto, C.; Morales, L.M.; Berrillo, S.; Vera Cuesta, H.; Baez, M.; Denis, M.; Marin, T.; Cabrera, Y.; et al. Non-invasive brain stimulation for children with autism spectrum disorders: A short-term outcome study. *Behav. Sci.* **2017**, *7*, 63. [CrossRef] [PubMed]

behavioral sciences

MDPI

Article

Non-Invasive Brain Stimulation for Children with Autism Spectrum Disorders: A Short-Term Outcome Study

Lázaro Gómez [1,4,*], Belkis Vidal [2], Carlos Maragoto [1,3], Lilia Maria Morales [4], Sheyla Berrillo [4], Héctor Vera Cuesta [1,3], Margarita Baez [4], Marlén Denis [1,3], Tairí Marín [1,3], Yaumara Cabrera [1,3], Abel Sánchez [1], Celia Alarcón [4], Maribel Selguera [2], Yaima Llanez [3], Lucila Dieguez [4] and María Robinson [5]

[1] Non Invasive Brain Stimulation Unit, International Center for Neurological Restoration, 25th Ave, Playa, Havana 15805, Cuba; maragoto@neuro.ciren.cu (C.M.); verac@neuro.ciren.cu (H.V.C.); mdenis@neuro.ciren.cu (M.D.); tairi@neuro.ciren.cu (T.M.); yaumara@neuro.ciren.cu (Y.C.); abel@neuro.ciren.cu (A.S.)
[2] Child and Adolescent Mental Health Service, Borrás-Marfán Hospital, G and 27 Street., Vedado, Havana 10400, Cuba; bvidalm@infomed.sld.cu (B.V.); marser@infomed.sld.cu (M.S.)
[3] Neuropediatric Clinic, International Center for Neurological Restoration, 25th Ave. Playa, Havana 15805, Cuba; yllanez@neuro.ciren.cu
[4] Clinical Neurophysiology Lab., International Center for Neurological Restoration, 25th Ave. Playa, Havana 15805, Cuba; lily@neuro.ciren.cu (L.M.M.); sheyla@neuro.ciren.cu (S.B.); minou@neuro.ciren.cu (M.B.); calarcon@neuro.ciren.cu (C.A.); lucila@neuro.ciren.cu (L.D.)
[5] Clinical Immunology Lab., International Center for Neurological Restoration, 25th Ave. Playa, Havana 15805, Cuba; robin@neuro.ciren.cu
* Correspondence: lazarog@neuro.ciren.cu; Tel.: +53-53-87-91-94

Received: 20 July 2017; Accepted: 14 September 2017; Published: 17 September 2017

Abstract: Non-Invasive Brain Stimulation (NIBS) is a relatively new therapeutic approach that has shown beneficial effects in Autism Spectrum Disorder (ASD). One question to be answered is how enduring its neuromodulatory effect could be. Twenty-four patients with ASD (mean age: 12.2 years) received 20 sessions of NIBS over the left dorsolateral prefrontal cortex (L-DLPFC). They were randomized into two groups with two (G1) or three (G2) clinical evaluations before NIBS. Both groups had a complete follow-up at six months after the intervention, with the aim of determining the short-term outcome using the total score on the Autism Behavior Checklist, Autism Treatment Evaluation Checklist, and the Autism Diagnostic Interview. Transcranial Direct Current Stimulation (tDCS) was used in ASD patients aged <11 years, and repetitive Transcranial Magnetic Stimulation (rTMS) for 11–13-year-olds. Observation points were at one, three, and six months after completing all the sessions of NIBS. A significant reduction in the total score on the three clinical scales was observed and maintained during the first six months after treatment, with a slight and non-significant tendency to increase the scores in the last evaluation. Twenty sessions of NIBS over the L-DLPFC improves autistic symptoms in ASD children, with a lasting effect of six months.

Keywords: autism; transcranial direct current stimulation; repetitive transcranial magnetic stimulation

1. Introduction

Autism Spectrum Disorders (ASD) still pose challenges for neuroscience for many reasons; one of these is how to apply the advances in neuroscience to understanding and treating ASD by improving diagnosis, interventions, and treatments. To date, diagnosis is made on clinical bases in the absence of a true biomarker for diagnosis and a better therapeutic approach. Different pharmacological and non-pharmacological therapies have shown positive results but are far from leading to a significant improvement in the core symptoms of patients with ASD. A high incidence of brain anomalies had been described by different methods in autistic patients in comparison with neurotypical patients, including MRI techniques and anatomic studies [1–3]. One interesting feature described by Casanova et al. was the abnormal structure in cortical mini-columns, with poor neuropil development in ASD patients, which probably accounts for intracortical inhibitory dysfunction [1,4]. Alterations in the GABAergic signaling pathway may characterize autistic neurobiology, which seems to not simply be related to a decreased GABA concentration, but probably to perturbations in key components of the GABAergic pathway beyond GABA levels, such as receptors and inhibitory neuronal density [5].

There is also some research exploring functional intracortical inhibition in motor areas of ASD patients that reinforce these ideas; nevertheless, its functional expression in terms of modulation by means of paired pulse stimulation protocol with TMS is really less remarkable than was expected, and it has been demonstrated in only a fraction of the ASD population [6]. In other experiments, indirect evidence has been described related to an aberrant plasticity based on an imbalanced excitatory and inhibitory cortical tone as well as biased GABAergic dysfunction [7].

Casanova et al. were the first to propose the use of Non-Invasive Brain Stimulation (NIBS) for ASD patients, based on the possibility that low-frequency repetitive TMS could somehow improve GABAergic neurotransmission; in their case, low-frequency repetitive Transcranial Magnetic Stimulation (rTMS) was used as a way to improve intracortical inhibition in ASD patients. They applied one weekly session over 12 weeks in a group of 25 ASD patients and described an increase in gamma activity in the EEG evoked by a visual processing paradigm, with changes in Event Related Potential (ERP), improved error monitoring, and correction function in a visual recognition task after the intervention [8,9]. There are just a few other studies providing evidence of the potential efficacy of both rTMS and transcranial direct current stimulation (tDCS) in ASD (excluding case reports). Unfortunately, most of the trials are limited to a low number of sessions (not more than 15 in the best-case scenario), and it was not known how enduring the modulatory effect was [10–13].

Dorsolateral prefrontal cortex (DLPFC) is the preferred target for many other clinical conditions, thus also in ASD based on its connections with other cortical and subcortical structures. Nevertheless, other considerations are valid about the feasibility for the selection of other targets [1,14]. Studies in ASD patients using low-frequency rTMS over the left DLPFC (L-DLPFC) have shown a significant improvement in some components of event-related cortical potentials (N200 and P300) as well as a significant reduction in response errors during cognitive tasks, repetitive behavior, and irritability [9]. Other authors using cathodal tDCS over the L-DLPFC have also reported beneficial effects for ASD in young adults. Curiously, there is also a published work of a trial with anodal stimulation over DLPFC reporting improvement in autistic symptomatology [10,11].

If we accept the hypothesis that an intra-cortical inhibitory dysfunction exists in ASD patients that can contribute to several dysfunctions in the motor, sensorial, and cognitive domains, we can expect some improvement in autistic behavior after applying brain-stimulating protocols that may potentiate intracortical inhibition, probably increasing the amount of available endogenous GABA in neuronal networks related to the stimulation target. In previous experiments, low-frequency rTMS and cathodal stimulation over the left DLPFC in children with ASD were associated with an improvement in autistic behavior one week after completing 20 sessions of transcranial stimulation [15,16]. One important question is how lasting this effect would be; in the present study, we describe the short-term clinical response during the first six months after treatment. We hypothesized that 20 sessions of NIBS over the L-DLPFC are expected to have a lasting positive neuromodulatory effect in patients with ASD.

2. Materials and Methods

2.1. Study Design

A controlled, randomized, and partial crossover trial was carried out in 24 children with ASD. Fifteen patients received the intervention after two clinical evaluations within one month (group 1, G1); nine patients started receiving the intervention after two months with three evaluation points at one-month intervals (group 2, G2). We assume that according to the stimulation protocol to be used (one daily session, up to a total of 20), it would be difficult to determine a washout period and patients from G2 ran in parallel to patients from G1 while they were receiving the intervention.

2.2. Sample Selection and Group Distribution

Children with ASD were recruited from the ambulatory service of the International Center for Neurological Restoration and Marfán-Borrás Hospital, from November 2015 to October 2016. Their diagnosis was made based on DSM-V diagnostic criteria, with the consensus of a multidisciplinary team including two Child Psychiatrists and two Neuropediatric Specialists. Only patients with a slight or moderate grade of severity were included according to their clinical characteristics, with stability during the last two months, and no changes in their therapeutic scheme in either pharmacologic or non-pharmacologic interventions. Diagnosis was confirmed by the results of the Childhood Autism Rating Scale (CARS) [17] and the Autism Diagnostic Interview, Revisited Edition (ADI-R, diagnostic algorithm) [18].

Patients with severe autistic behavior were excluded, as well as patients with concurrent epilepsy. Patients with unclear diagnosis or with diagnostic disagreement between neurologist and psychiatrist opinion were also excluded. If any change was needed in their therapeutic scheme, the patient was also excluded from the trial. We applied a simple randomization technique to assign patients to G1 (early intervention group, after one month of follow-up) or G2 (intervention after two months of follow-up).

2.3. Clinical Evaluation

Three main clinical scales were used as outcome measurements for all patients, according to their parent opinions: ADI-R (algorithm for current condition), the Autism Behavioral Checklist (ABC) [19], and the Autism Treatment Evaluation Checklist (ATEC) [20]. All the scales were applied by a Child Psychiatrist who was not involved in the intervention, and patients were evaluated at one, three, and six months after completing the 20 sessions. The Global Clinical Impression Scale (GCIS) [21] was also applied as a complementary, more qualitative evaluation. The Wilcoxon matched pair test and Mann–Whitney U test were used either for intra- or between-group comparisons at the different evaluation moments ($\alpha = 0.05$).

2.4. Neurophysiological Evaluation

2.4.1. Functional Brain Connectivity

An electroencephalogram (EEG) based connectivity analysis was carried out in all the ASD patients, but only 15 good-quality EEG recording were obtained for connectivity analysis (nine from the tDCS group and six from the rTMS group). EEG traces taken one week before starting the intervention and one week after finishing it were analyzed. Significant windows were selected from the EEG trace in open eyes state (38 from each patient), because they were more often artifact-free than the closed-eyes state.

We used a 19-electrode montage including Fp1, Fp2, F7, F8, F3, F4, C3, C4, T5, T6, T3, T4, P3, P4, O1, O2, Fz, Cz, and Pz, from the 10/20 system. Electrode Impedance was kept below 5 kΩ. Functional connectivity was analyzed based on the synchronization likelihood between electrodes for the five frequency bands: δ (1–3.9 Hz), θ (4–7.9 Hz) α (8–12.9 Hz), β (13–29.9 Hz), and γ (30–35 Hz) [22]. Mathematical analysis was developed with connectivity algorithms implemented in MATLAB v.7.7

R2008b. Significant connectivity ($p < 0.05$) in each frequency band was represented on an X/Y coordinate according to the 10/20 system for each brain stimulation modality independently, and the results by group were represented over the scalp surface map.

2.4.2. Event-Related Potentials

Only in six children was it possible to carry out a passive oddball paradigm for P300 Event-Related Potentials (ERPs) before the intervention; four of them received tDCS and two rTMS. Patients were seated in a sound- and light-attenuated room while a paradigm was conducted consisting of 200 stimuli, 80% frequent (500 Hz) and 20% infrequent target (1000 Hz) tones, delivered through headphones while children were watching a silent movie. In the passive version of the oddball task the subject´s attention is usually directed away from the sequence of standard and deviant tones toward another, moderately demanding task, usually in a different modality [23]. We proposed a version of this paradigm considering the characteristics of the autistic children, with poor collaboration. All stimuli (50 ms; 5 ms rise and fall time) were presented binaurally with an inter-stimulus interval of 1300 ms.

To record the auditory P300 ERPs, 19 surface electrodes (Ag/ClAg) were attached to the scalp according to the international electrode placement 10–20 system (Fp1, Fp2, F3, F4, F7, F8, C3, C4, T3, T4, T5, T6, P3, P4, O1, O2, Fz, Cz, Pz), with additional electrodes for EOG (lateral to the outer canthus of the right eye and above the middle of the left eyebrow), and referenced to the linked earlobe. The electrode impedance was kept below 5 kΩ. EEG data were off-line averaged (100 ms prior and 700 ms after stimulus onset), and P300 latency and amplitude were quantified at Fz, Cz, and Pz. A continuous acquisition system was employed (Medicid 5, Neuronic SA, Cuba) and EEG data were EOG-corrected offline. The sampling rate of all channels was 200 Hz. To assure auditory system normality, Auditory Brainstem Response was previously recorded. The stimuli were 0.1 ms alternating clicks delivered through a headphone (DR-531B-7, Elegas Acous Co. Ltd., Tokyo, Japan). The records were obtained using the evoked potentials measuring system Neuropack M1 (Nihon Kohden, Tokyo, Japan).

P300 ERPs were obtained before and after NIBS, and P300 latency and amplitude values from the two evaluations were compared (from Fz electrode position; Wilcoxon matched pairs test, $\alpha < 0.05$). All measurements were of the difference wave (infrequent wave minus frequent wave) and in this case, due to the low number of patients' recordable P300, the analysis was performed considering all patients as a single group.

2.5. Intervention

Patients received one daily session of NIBS, from Monday to Friday, for a total of 20 sessions. tDCS was used in patients 10 years old or younger, and rTMS in patients 11 and older. The reason for using tDCS in younger children instead of rTMS was that in this case a major degree of collaboration is needed to assure effective focal stimulation over the selected target (left dorsolateral prefrontal cortex, L-DLPFC). During the stimulation sessions patients, were comfortably seated while watching TV cartoons of their preference or listening to music and playing with small, simple toys.

2.5.1. Transcranial Direct Current Stimulation (tDCS)

tDCS was only used in patients 10 years old or younger (Neuroconn tDCS stimulator, München, Germany). A cathode was positioned over F3 (10/20 international EEG electrode system), and an anode over the proximal right arm. The stimulation intensity was 1 mA, and was maintained for 20 min. Each electrode pad was humidified with a 0.9% NaCl solution.

2.5.2. Repetitive Transcranial Magnetic Stimulation (rTMS)

rTMS was used in patients older than 10 years and 11 months (MagStim Rapid^2, Whitland, UK). A butterfly coil with air cooling system (MagStim Double 70 mm Air Film Coil) was used. The center of the coil was located over F3, back handed 45° from the midline, and a total of 1500 pulses were delivered in each session, at 1 Hz of frequency and an intensity of 90% of the resting motor threshold (i.e., the minimum stimulation intensity required to elicit a discernible hand muscle response in at least three of five consecutive pulses) [24]. Sessions were subdivided into four trains of 375 pulses, with a 1-min interval between each allowing for free movement of the head and neck. Most of the patients rejected the use of earplugs, and in the best-case scenario they agreed to listen to music by wearing headphones from their own devices.

2.6. Ethical Considerations

All the procedures followed the rules of the Declaration of Helsinki of 1975 for human research, and the study was approved by the scientific and ethics committee from the International Center for Neurological Restoration (CIREN37/2015). Parents gave written informed consent for their children to be considered for inclusion in the study, and where possible children also gave their consent.

3. Results

3.1. Change in Clinical Scales One Month after the Intervention

As a global result (G1 + G2), a significant decrease in the total score was observed in ADI-R, ABC, and ATEC scales one month after the intervention (Wilcoxon matched pair test; ABC, $Z = 3.823$, $p = 0.000131$; ADI-R, 3.337, $p = 0.000846$; ATEC, $Z = 3.723$, $p = 0.000196$). Figure 1a represents the global scores for the three main clinical scales we applied for the evaluation of autism behavior. The values are in correspondence with qualitative changes described by their relative uniqueness in socialization and communication domains, and with the GCIS (pre-intervention: 3.47 ± 0.6; post-intervention: 2.95 ± 0.2. Wilcoxon pair series test: $Z = 2.803$, $p = 0.005062$). All of the patients' autistic behavior improved according to the scale results observed one month after the intervention. A comparison between the change in the total score of clinical scales did not show any significant differences correlated with the use of tDCS and rTMS (Mann–Whitney U Test, $p > 0.05$). Initial evaluation did not show any differences in the total score between groups when we looked for any evidence of age-dependent characteristics in both groups. The clinical response to NIBS was apparently independent of group age and type of intervention (See Figure 1b).

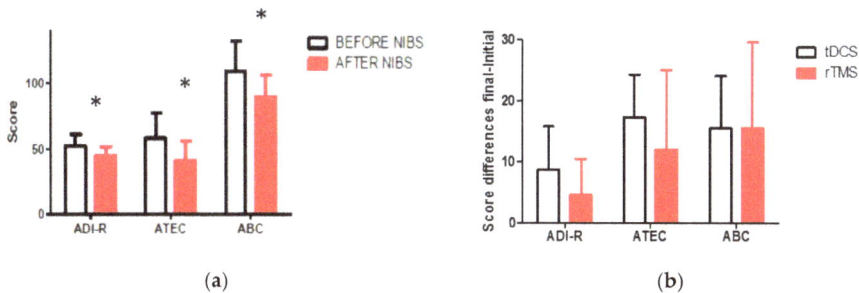

Figure 1. (a) Comparison of pre- intervention and one-month post-intervention scores in clinical scales; (b) no significant differences in the clinical effect with the use of rTMS or tDCS according to the score differences (Initial–Final) (* $p < 0.05$).

Figure 2 shows that no significant differences were observed between G1 and G2 in the initial clinical scale scores before the intervention (Mann–Whitney U test, $p > 0.05$); but values observed in G1 after NIBS showed a very important change when compared with the initial values (Wilcoxon; ABC, $Z = 3.823$, $p = 0.000132$; ADI-R, $Z = 3.550$, $p = 0.000385$;ATEC, $Z = 5.526$, $p = 0.000241$); the same behavior was seen in G2 after the intervention (Wilcoxon; ABC, $Z = 3.295$, $p = 0.000982$; ADI-R, $Z = 3.588$, $p = 0.009633$; ATEC, $Z = 3.179$, $p = 0.001474$).

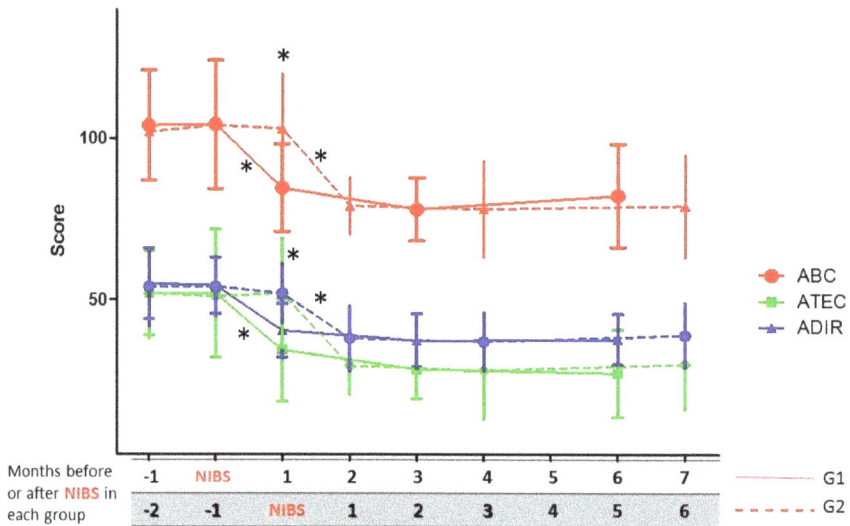

Figure 2. Short-term outcome based on clinical scale scores until six months of follow-up after NIBS in group 1 (G1) and group 2 (G2). * $p < 0.05$ (significant differences; comparisons between and within each group with scale scores before and after NIBS).

3.2. Change in Clinical Scales during the First Six Months after the Intervention

Post-intervention follow-up was extended over six months. For both groups G1 and G2, NIBS induced a significant change in clinical scale scores (See Figure 2), and both groups maintained approximately the same values until the sixth month after NIBS, when there was a slight tendency for the total score in ABC, ADI-R, and ATEC to increase, but their punctuation remained lower than the observed values before treatment (See Figure 2). According to their parents' opinion, stereotypes intensity and variety increased in frequency and number in many patients around the sixth month, but in clinical scales there were not any significant changes in specific domains.

Group 1 had lower values for all the scales one month after NIBS (solid lines); the same happened with G2 patients running in parallel, but only since the fourth month and ahead of follow-up (also one month after NIBS; dashed lines).

3.3. EEG-Based Brain Functional Connectivity Analysis

An increase in brain functional connectivity was observed after NIBS for patients who received either tDCS or rTMS, especially for higher frequencies: α, β, and γ bands showed the greatest increase, with a large increment in the number of 10/20 system points functionally related. The analysis of the effect in functional connectivity showed that the group that received rTMS accounted for the most significant changes when initial and final values were compared (see Figure 3).

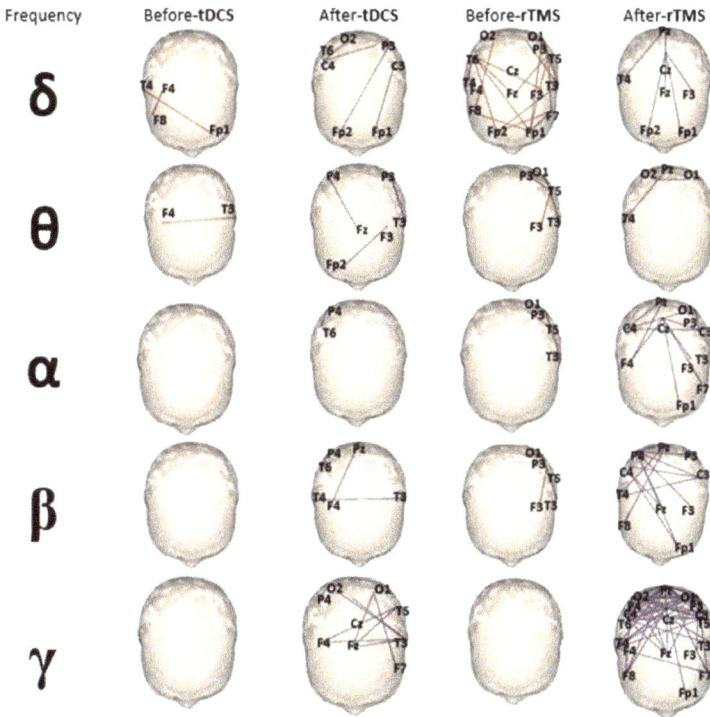

Figure 3. EEG-based analysis of brain functional connectivity. Note the positive change in all frequency bands after the intervention, but mainly in the γ band for both tDCS and rTMS, especially in the group that received rTMS (all electrodes' locations are statistically significant, $p < 0.05$).

3.4. ERP Analysis

Recordings were possible in only six patients in the initial evaluation; thus, after the intervention, only activity in those patients who collaborated during the task was recorded. All patients had a normal auditory brainstem response according to the normative laboratory data, but one patient had no ERP response. The P300 component showed a scalp distribution in the frontal and central regions, maximal in Fz. There was a shortening of P300 latency after treatment, with statistically significant differences with respect to the basal response (Wilcoxon matched pairs test, $p = 0.043$). P300 amplitude did not show statistically significant differences after treatment, even though the average value in the group was higher after the intervention (see Figure 4). The prolonged latency and low amplitude may be in agreement with the abnormal connectivity of the frontal lobes seen in ASD subjects [25], and engaged with attention circuits. The improvement of these parameters after NIBS might be a consequence of the improvement in functional connectivity and probably correlated with positive clinical changes in attentiveness, concentration, and speed of answering questions from a parent.

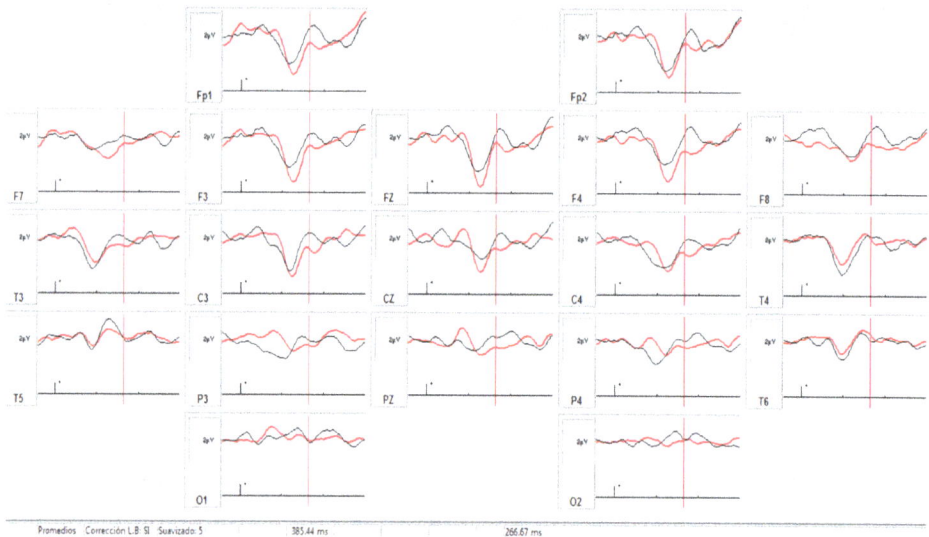

Figure 4. Grand averaged infrequent target. Black trace: pre-intervention (Amplitude and latency at Fz: 3.09 μV, 410 ms); red trace: post-intervention (Amplitude and latency at Fz: 3.28 μV, 371 ms).

4. Discussion

Our results reproduce similar neuromodulatory effects to those reported previously by two other research groups who used the same target (L-DLPFC), applying either rTMS or tDCS [9,11,26]. Casanova et al. pioneered the use of therapeutic NIBS in autism and described the effects of one weekly low-frequency rTMS (1 Hz) during 12 weeks in 25 ASD children and adolescents. In that study they reported a significant reduction in repetitive and restricted behavior patterns and irritability, but no changes in social awareness or hyperactivity. They also described an increase in the amount of gamma activity in the EEG, shortening of the P300 component of ERP, and an increase of amplitude in the group receiving the intervention [9]. They used 150 TMS pulses per session, one session per week, and 12 sessions in total. In our study, the number of stimuli per session and the total number of sessions were considerably higher; this is in correspondence with recommended practical aspects for conventional use of NIBS in other diseases such as depression, obsessive compulsive disorder, chronic pain, etc., in order to obtain a sustained effect [27,28]. This could likely explain the significant clinical improvement in our group, not only in the abovementioned domains, but also in social aspects and communication. We also were able to establish that NIBS' effects persisted into the sixth month after completion of treatment, a result that, as far as we know, has never been reported by any other group; this is a very important finding because it would probably indicate that patients would need a new treatment cycle (e.g., a second cycle of NIBS) six months after the previous intervention to maintain clinical improvement. Of course, it would also be interesting to know how the behavior of our patients will develop over the next six months and the forthcoming years (long-term outcome).

Another interesting double-blind and randomized trial in adult ASD patients was published by Enticott et al. using the dorsomedial prefrontal cortex as the target, and stimulating with the HAUT coil, which makes it possible to reach deep structures in the brain [12]. They applied 10 daily sessions of 5 Hz, 1500 pulses at resting motor threshold intensity. The clinical response was evaluated immediately after the intervention and one month later, and they reported a significant reduction in social relating symptoms (relative to sham participants) for both post-treatment assessments. Those in the active condition also showed a reduction in self-oriented anxiety during difficult and emotional social situations from pre-treatment to one-month follow-up. They suggested the need to use extended protocols such as those used to treat depression, and we agree that it is essential to achieving lasting effects for both NIBS methods, rTMS or tDCS [28].

D'Urso et al. published the results of an open pilot study using cathodic tDCS over the L-DLPFC in 10 young adults (18–25 years old) with 10 daily sessions of 1.5 mA, with evaluation two weeks after the intervention. They reported an improvement in autistic behavior of 26.7% according to the change observed in the clinical scale they applied; only two patients did not change their initial scale values, while the rest of the group improved in their symptoms [11]. There are two main differences in the tDCS protocol we applied in younger children (5–10 years old): first, again we used a higher number of sessions and second, the stimulation intensity in our case was 1 mA. NIBS in children seems to be as safe as in the adult population, so the same safety guidelines are indicated for its use in any population [29,30]. It is important to mention that in the case of tDCS, there is not enough evidence about the safety of 2 mA in young people (<18 years old), and most research has been done using 1 mA; however, there are also a few papers describing the absence of adverse effects with the use of 2 mA in pediatric patients [13].

In another interesting article, Amatachaya et al. described the effects of anodal tDCS over L-DLPF, with improvement in autistic behavior in 20 children with ASD. They designed a randomized, double-blind crossover and sham-controlled trial, applying 1 mA anodal tDCS over five consecutive days, and a washout period of four weeks This result reactivates the essential research question as to whether the clinical, biochemical, and physiological effects are polarity-dependent or not [10,31–35].

There are results from clinical trials with other diseases and physiological studies in which the opposite effects of cathodic and anodic tDCS have been well documented [35,36]. Our results using cathodic tDCS or low-frequency rTMS over the L-DLPFC share the same theoretical basis developed by Casanova et al. and D'Urso et al., though we studied a low sample number (24 patients), and by design we employed an open trial without the use of placebo stimulation. Nevertheless, the results are reproducible in G1 and G2, with patients being their own controls from the start of the intervention. We also had the opportunity to compare the effects of low-frequency rTMS and cathodic tDCS at the same time, but in ASD patients of different ages and with our experimental design we demonstrated that there are no differences, at least not in their clinical effects. Certainly, there was an important effect achieved by increasing functional connectivity with the use of both the stimulating methods, rTMS and tDCS, with a clear superiority of rTMS-induced changes. Of course, more research is needed to draw well-documented conclusions. Other authors have described improvements in particular aspects of adult ASD patients such as working memory and better syntax acquisition following single-dose anodal tDCS over the left and right dorsolateral prefrontal cortex (first case) or exclusively over the left dorsolateral prefrontal cortex [13,37] There were also positive changes in P300 ERP, but due to the low number of subjects in whom it was possible to carry out the passive oddball paradigm, we cannot make any comparison between the two different stimulating methods. Unfortunately, sample sizes for EEG-based connectivity analysis and ERP paradigm could not include all the patients, because, in close relationship with the complexity of the used method, when we carried out their initial electrophysiological evaluation (before intervention), not all recordings had a high enough quality for analysis.

5. Conclusions

Twenty sessions of NIBS over L-DLPFC improves autistic symptoms in ASD children, with a lasting effect of six months.

Acknowledgments: The authors appreciate the contribution of all parents of and children participating in the study. We also thank Tlalli Mota and Jim Clifford for helping with the review of the manuscript.

Author Contributions: All authors contributed substantially to the work reported. Lázaro Gómez participated in the design of the research, NIBS sessions, analysis, and discussion of the results, and wrote the manuscript. Belkis Vidal and Maribel Serguera participated in the diagnosis of patients and clinical evaluation of all the scales. Hector Vera Cuesta and Carlos Maragoto Rizo contributed to the diagnosis and clinical data of patients. Lilia Maria Morales and Sheyla Berrillo contributed to the analysis of the EEG traces and functional connectivity calculations. Margarita Báez carried out ERP recordings and analysis. Abel Sánchez and Celia Alarcón performed the technical work for recordings of EEG and ERP. Tairí Marín, Marlén Denis, Yaumara Cabrera, Yaima LLanez, and Lucila Duiéguez participated in the NIBS sessions. Maria de los Angeles Robinson Agramonte participated in the analysis and discussion of the results and review of the manuscript.

Conflicts of Interest: The authors declare no conflict of interest.

References

1. Casanova, M.F.; Buxhoeveden, D.P.; Switala, A.E.; Roy, E. Minicolumnar pathology in autism. *Neurology* **2002**, *58*, 428–432. [CrossRef] [PubMed]
2. Freitag, C.M.; Luders, E.; Hulst, H.E.; Narr, K.L.; Thompson, P.M.; Toga, A.W.; Krick, C.; Konrad, C. Total brain volume and corpus callosum size in medication-naive adolescents and young adults with autism spectrum disorder. *Biol. Psychiatry* **2009**, *66*, 316–319. [CrossRef] [PubMed]
3. Kumar, A.; Sundaram, S.K.; Sivaswamy, L.; Behen, M.E.; Makki, M.I.; Ager, J.; Janisse, J.; Chugani, H.T.; Chugani, D.C. Alterations in frontal lobe tracts and corpus callosum in young children with autism spectrum disorder. *Cereb. Cortex* **2010**, *20*, 2103–2113. [CrossRef] [PubMed]
4. Casanova, M.F.; El-Baz, A.; Vanbogaert, E.; Narahari, P.; Switala, A. A topographic study of minicolumnar core width by lamina comparison between autistic subjects and controls: Possible minicolumnar disruption due to an anatomical element in-common to multiple laminae. *Brain Pathol.* **2010**, *20*, 451–458. [CrossRef] [PubMed]
5. Fatemi, S.H.; Reutiman, T.J.; Folsom, T.D.; Rustan, O.G.; Rooney, R.J.; Thuras, P.D. Downregulation of GABAA receptor protein subunits alpha6, beta2, delta, epsilon, gamma2, theta, and rho2 in superior frontal cortex of subjects with autism. *J. Autism Dev. Disord.* **2014**, *44*, 1833–1845. [CrossRef] [PubMed]
6. Enticott, P.G.; Kennedy, H.A.; Rinehart, N.J.; Tonge, B.J.; Bradshaw, J.L.; Fitzgerald, P.B. GABAergic activity in autism spectrum disorders: An investigation of cortical inhibition via transcranial magnetic stimulation. *Neuropharmacology* **2013**, *68*, 202–209. [CrossRef] [PubMed]
7. Oberman, L.M.; Pascual-Leone, A.; Rotenberg, A. Modulation of corticospinal excitability by transcranial magnetic stimulation in children and adolescents with autism spectrum disorder. *Front. Hum Neurosci.* **2014**, *8*. [CrossRef] [PubMed]
8. Baruth, J.M.; Casanova, M.F.; El-Baz, A.; Horrell, T.; Mathai, G.; Sears, L.; Sokhadze, E. Low-frequency repetitive transcranial magnetic stimulation (rTMS) modulates evoked-gamma frequency oscillations in autism spectrum disorder (ASD). *J. Neurother.* **2010**, *14*, 179–194. [CrossRef] [PubMed]
9. Casanova, M.F.; Baruth, J.M.; El-Baz, A.; Tasman, A.; Sears, L.; Sokhadze, E. Repetitive transcranial magnetic stimulation (rTMS) modulates event-related potential (ERP) indices of attention in autism. *Transl. Neurosci.* **2012**, *3*, 170–180. [CrossRef] [PubMed]
10. Amatachaya, A.; Auvichayapat, N.; Patjanasoontorn, N.; Suphakunpinyo, C.; Ngernyam, N.; Aree-uea, B.; Keeratitanont, K.; Auvichayapat, P. Effect of anodal transcranial direct current stimulation on autism: A randomized double-blind crossover trial. *Behav. Neurol.* **2014**, *2014*. [CrossRef] [PubMed]
11. D'Urso, G.; Bruzzese, D.; Ferrucci, R.; Priori, A.; Pascotto, A.; Galderisi, S.; Altamura, A.C.; Bravaccio, C. Transcranial direct current stimulation for hyperactivity and noncompliance in autistic disorder. *World J. Biol. Psychiatry* **2015**, *16*, 361–366. [CrossRef] [PubMed]
12. Enticott, P.G.; Fitzgibbon, B.M.; Kennedy, H.A.; Arnold, S.L.; Elliot, D.; Peachey, A.; Zangen, A.; Fitzgerald, P.B. A double-blind, randomized trial of deep repetitive transcranial magnetic stimulation (rTMS) for autism spectrum disorder. *Brain Stimul.* **2014**, *7*, 206–211. [CrossRef] [PubMed]
13. Schneider, H.D.; Hopp, J.P. The use of the Bilingual Aphasia Test for assessment and transcranial direct current stimulation to modulate language acquisition in minimally verbal children with autism. *Clin. Linguist. Phon.* **2011**, *25*, 640–654. [CrossRef] [PubMed]
14. Downar, J.; Daskalakis, Z.J. New targets for rTMS in depression: A review of convergent evidence. *Brain Stimul.* **2013**, *6*, 231–240. [CrossRef] [PubMed]

15. Gómez, L.; Denis, M.; Marín, T.; Vidal, B.; Maragoto, C.; Vera, C.; Serguera, M.; Morales, L.; Báez, M.; Sánchez, A.; et al. Estudio piloto sobre el efecto de la Estimulación Cerebral No Invasiva en el Trastorno del Espectro Autista. *Rev. Mex. Neurocienc.* **2016**, *17*, 51.

16. Gómez, L.; Denis, M.; Marín, T.; Vidal, B.; Maragoto, C.; Vera, H.; Serguera, M.; Morales, L.; Báez, M.; Sánchez, A.; et al. Non invasive brain stimulation in children with autism spectrum disorder. *Brain Stimul.* **2017**, *10*, 347. [CrossRef]

17. Schopler, E.; Reichler, R.; DeVellis, R.; Daly, K. Toward objective classification of childhood autism: Childhood Autism Rating Scale (CARS). *J. Autism Dev. Disord.* **1980**, *10*, 91–103. [CrossRef] [PubMed]

18. Lord, C.; Rutter, M.; Le Couteur, A. Autism diagnostic interview-revised: A revised version of a diagnostic interview for caregivers of individuals with possible pervasive developmental disorders. *J. Autism Dev. Disord.* **1994**, *24*, 659–685. [CrossRef] [PubMed]

19. Krug, D.; Arisk, J.; Almond, P. Behavior checklist for identifying severely handicapped individuals with high levels of autistic behavior. *J. Child Psychol. Psychiatry* **1980**, *21*, 221–229. [CrossRef] [PubMed]

20. Magiati, I.; Moss, J.; Yates, R.; Charman, T.; Howlin, P. Is the Autism Treatment Evaluation Checklist a useful tool for monitoring progress in children with autism spectrum disorders? *J. Intellect. Disabil. Res.* **2011**, *55*, 302–312. [CrossRef] [PubMed]

21. Busner, J.; Targum, S. The clinical global impressions scale: Applying a research tool in clinical practice. *Psychiatry* **2007**, *4*, 28–37. [PubMed]

22. Stam, J.C.; van Dijk, B.W. Synchronization likelihood: An unbiased measure of generalized synchronization in multivariate data sets. *Phys. D Nonlinear Phenom.* **2002**, *163*, 236–251. [CrossRef]

23. Näätänen, R. The role of attention in auditory information processing as revealed by event-related potentials and other brain measures of cognitive function. *Behav. Brain Sci.* **1990**, *13*, 201–233. [CrossRef]

24. Rossi, S.; Hallet, M.; Rossini, P.M.; Pascual-Leone, A. Safety, ethical considerations, and application guidelines for the use of transcranial magnetic stimulation in clinical practice and research. *Clin. Neurophysiol.* **2009**, *120*, 2008–2039. [CrossRef] [PubMed]

25. Catani, M.; Dell'Acqua, F.; Budisavljevic, S.; Howells, H.; Thiebaut de Schotten, M.; Froudist-Walsh, S.; D'Anna, L.; Thompson, A.; Sandrone, S.; Bullmore, E.; et al. Frontal networks in adults with autism spectrum disorder. *Brain* **2016**, *139*, 616–630. [CrossRef] [PubMed]

26. Baruth, J.M.; Wall, C.A.; Patterson, M.C.; Port, J.D. Proton magnetic resonance spectroscopy as a probe into the pathophysiology of autism spectrum disorders (ASD): A review. *Autism Res.* **2013**, *6*, 119–133. [CrossRef] [PubMed]

27. Lefaucheur, J.P.; Andre-Obadia, N.; Poulet, E.; Devanne, H.; Haffen, E.; Londero, A.; Cretin, B.; Leroi, A.M.; Radtchenko, A.; Saba, G.; et al. French guidelines on the use of repetitive transcranial magnetic stimulation (rTMS): Safety and therapeutic indications. *Neurophysiol. Clin.* **2011**, *41*, 221–295. [CrossRef] [PubMed]

28. Perera, T.; George, M.; Grammer, G.; Janicak, P.; Pascual-Leone, A.; Wirecki, T. The clinical TMS society consensus review and treatment recommendations for TMS therapy for major depressive disorder. *Brain Stimul.* **2016**, *9*, 336–346. [CrossRef] [PubMed]

29. Hameed, M.Q.; Dhamne, S.C.; Gersner, R.; Kaye, H.L.; Oberman, L.M.; Pascual-Leone, A.; Rotenberg, A. Transcranial magnetic and direct current stimulation in children. *Curr. Neurol. Neurosci. Rep.* **2017**, *17*. [CrossRef] [PubMed]

30. Krishnan, C.; Santos, L.; Peterson, M.; Ehinger, M. Safety of noninvasive brain stimulation in children and adolescents. *Brain Stimul.* **2015**, *8*, 76–87. [CrossRef] [PubMed]

31. Jacobson, L.; Koslowsky, M.; Lavidor, M. tDCS polarity effects in motor and cognitive domains: A meta-analytical review. *Exp. Brain Res.* **2012**, *216*, 1–10. [CrossRef] [PubMed]

32. Joos, K.; De, R.D.; Van de Heyning, P.; Vanneste, S. Polarity specific suppression effects of transcranial direct current stimulation for tinnitus. *Neural Plast.* **2014**, *2014*. [CrossRef] [PubMed]

33. Ladeira, A.; Fregni, F.; Campanha, C.; Valasek, C.A.; De, R.D.; Brunoni, A.R.; Boggio, P.S. Polarity-dependent transcranial direct current stimulation effects on central auditory processing. *PLoS ONE* **2011**, *6*, e25399. [CrossRef] [PubMed]

34. Sohn, M.K.; Jee, S.J.; Kim, Y.W. Effect of transcranial direct current stimulation on postural stability and lower extremity strength in hemiplegic stroke patients. *Ann. Rehabil. Med.* **2013**, *37*, 759–765. [CrossRef] [PubMed]

35. Stagg, C.J.; Best, J.G.; Stephenson, M.C.; O'Shea, J.; Wylezinska, M.; Kincses, Z.T.; Morris, P.G.; Matthews, P.M.; Johansen-Berg, H. Polarity-sensitive modulation of cortical neurotransmitters by transcranial stimulation. *J. Neurosci.* **2009**, *29*, 5202–5206. [CrossRef] [PubMed]

36. D'Urso, G.; Brunoni, A.R.; Anastasia, A.; Micillo, M.; Mantovani, A. Polarity-dependent effects of transcranial direct current stimulation in obsessive-compulsive disorder. *Neurocase* **2016**, *22*, 60–64. [CrossRef] [PubMed]

37. Van Steenburgh, J.J.; Varvaris, M.; Schretlen, J.; Vannorsdall, T.J.; Gordon, B. Balanced bifrontal transcranial direct current stimulation enhaces working memory in adults with high-functioning autism: A sham-controlled crossover study. *Mol. Autism* **2017**, *8*, 40. [CrossRef] [PubMed]

behavioral
sciences

MDPI

Article

Atypical Processing of Novel Distracters in a Visual Oddball Task in Autism Spectrum Disorder

Estate M. Sokhadze [1,2], **Eva V. Lamina** [1], **Emily L. Casanova** [1,2], **Desmond P. Kelly** [1,2], **Ioan Opris** [3], **Irma Khachidze** [4] and **Manuel F. Casanova** [1,2,*]

1   Department of Biomedical Sciences, University of South Carolina School of Medicine-Greenville, 200 Patewood Dr., Ste A200, Greenville, SC 29615, USA; SOKHADZE@greenvillemed.sc.edu (E.M.S.); LAMINAE@mailbox.sc.edu (E.V.L.); ECasanova@ghs.org (E.L.C.); DKelly@ghs.org (D.P.K.)
2   Developmental Behavioral Unit, Department of Pediatrics, Children's Hospital, Greenville Health System, Greenville, SC 29615, USA
3   School of Medicine, University of Miami, Miami, FL 33136, USA; ioanopris.phd@gmail.com
4   Centre of Experimental Biomedicine, 14 Gotya str., Tbilisi 0160, Georgia; irmakha@yahoo.com
*   Correspondence: MCasanova@ghs.org; Tel.: +1-(864)-454-4585

Received: 24 September 2017; Accepted: 14 November 2017; Published: 16 November 2017

Abstract: Several studies have shown that children with autism spectrum disorder (ASD) show abnormalities in P3b to targets in standard oddball tasks. The present study employed a three-stimulus visual oddball task with novel distracters that analyzed event-related potentials (ERP) to both target and non-target items at frontal and parietal sites. The task tested the hypothesis that children with autism are abnormally orienting attention to distracters probably due to impaired habituation to novelty. We predicted a lower selectivity in early ERPs to target, frequent non-target, and rare distracters. We also expected delayed late ERPs in autism. The study enrolled 32 ASD and 24 typically developing (TD) children. Reaction time (RT) and accuracy were analyzed as behavioral measures, while ERPs were recorded with a dense-array EEG system. Children with ASD showed higher error rate without normative post-error RT slowing and had lower error-related negativity. Parietal P1, frontal N1, as well as P3a and P3b components were higher to novels in ASD. Augmented exogenous ERPs suggest low selectivity in pre-processing of stimuli resulting in their excessive processing at later stages. The results suggest an impaired habituation to unattended stimuli that incurs a high load at the later stages of perceptual and cognitive processing and response selection when novel distracter stimuli are differentiated from targets.

Keywords: event-related potential; autism spectrum disorder; attention; cognitive processes; reaction time

1. Introduction

Autism Spectrum Disorder (ASD) is characterized by severe disturbances in reciprocal social relations, varying degrees of language and communication difficulties, and behavioral patterns which are restricted, repetitive, and stereotyped [1]. Additionally, individuals with autism usually present excessive reactions to change in their environment such as aversive reactions to visual, auditory, and tactile stimuli. These perception and sensory reactivity abnormalities, found in a majority of subjects with ASD, affect their ability to effectively process information [2]. In a series of electrophysiological studies conducted by our group we explored specifics of event-related potential (i.e., ERP) reflecting information processing during performance on reaction time (RT) tasks in children with ASD [3–9]. Our prior studies explored the manifestations of excessive local connectivity and impaired distal functional connectivity, excessive cortical excitation/inhibition ratio, and deficient executive functioning in ASD by analyzing behavioral performance on attention tasks with concurrent dense-array ERP recording. More detailed theoretical considerations related to the results of these studies were discussed in our reviews on this topic [9–11].

The current study explored atypicality of reactivity to novel stimuli in autism during performance on a three-stimulus visual oddball task as reflected in RT and ERP. Elicitation of ERP waves related to novelty processing can be readily achieved through this type of oddball paradigm, in which subjects are exposed to continuous succession of three types of stimuli, one presented frequently (standard), while the two other types of stimuli are rare. One of the rare stimuli is designated as a target, whereas the second type rare stimuli has some distinction from target and is usually referred to as a novel, or novel distracter. Analysis of ERP components using this test paradigm is a useful way of investigating task-relevant and -irrelevant information processing stages as well as selective attention. Amplitude and latency characteristics of negative and positive ERP components at selected scalp topographic regions-of-interest (ROI) can provide valuable information about the early sensory perception processes and the higher-level processes including attention, cortical inhibition, response selection, error monitoring, memory update, and other cognitive activity related to working memory [11–15].

Event-related potentials reflect the activation of neural structures in primary sensory cortex, and in associative cortical areas related to higher order cognitive processes. ERP components can be categorized as short-latency (exogenous, e.g., N1) or long-latency (endogenous, e.g., P3) ERPs, which reflect early-stage, modality-specific and late-stage polymodal associative processing respectively. The early ERP components (e.g., P1, N1) reflect exogenous processes modulated by the physical attributes of the stimulus (i.e., brightness for visual stimuli; loudness of auditory stimuli, etc.), rather than by endogenous cognitive processes [16]. However, it has been noted that attention processes may operate even at the early stages of information intake and influence stimulus processing at the later stage [17]. Two endogenous components of the ERP, namely the N2 and P3, are thought to be directly associated with the cognitive processes of perception and selective attention [18]. The posterior visual N2 (labeled as N2b) is enhanced if the presented stimulus contains a perceptual feature or attributes defining the target in the task. In majority of oddball tasks, the N2b is related to the cognitive processes of stimulus identification and distinction [18,19]. In a modification of the oddball task, when rare distracters are presented along with standard and rare target stimuli, these distracters elicit a fronto-central P3a, whereas the targets elicit a parietally distributed P3b [20,21]. In a three stimulus oddball task the P3a is interpreted as "orienting" of attention to novelty, and the P3b as an index of ability to sustain attention to target. Error sensitivity can be examined by measuring response-locked ERP components associated with cortical responses to committed errors. Two ERP components relevant in this context are the error-related negativity (ERN) and the error-related positivity (Pe).

This study employed a visual novelty oddball task with simultaneous recording of motor responses and brain potentials in children with ASD and in typically developing (TD) children. Such an approach allowed us to analyze attentional and cognitive processing mechanisms recruited in typically developing subjects relative to children with autism. We proposed that early stages of the sensory visual signal processing in autism might be characterized by low selectivity and by reduced adaptation to task-irrelevant items. We predicted that in autism, as compared to TD controls, there would be a lower selectivity in early ERPs components in response to infrequent target, frequent non-target and rare novel distracters as well as delayed endogenous ERP components. In particular, we anticipated higher amplitudes of P1 and N1 not only to target but also to standard and novel stimuli reflective of deficits in stimulus category selectivity. Amplitude and latency of N2 and P3 components in ASD and TD children were expected to differ during later stages of information processing suggesting that a more effortful allocation was needed to sustain attention on target differentiation from the novel distracters. Differentiation of motor responses in terms of their correctness reflected in error-related potentials (ERN, Pe) was also expected to be deficient in children with ASD, and was predicted to negatively affect accuracy of motor responses.

2. Methods and Materials

2.1. Participants

Participants with autism spectrum disorder (ASD) (age range 9 to 18 years) were recruited through the Weisskopf Child Evaluation Center (WCEC). Diagnosis was made according to the Diagnostic and Statistical Manual of Mental Disorders (DSM-IV-TR) [22] or DSM-5 [1] and further ascertained with the Autism Diagnostic Interview—Revised (ADI-R) [23]. They also had a medical evaluation by a developmental pediatrician. All subjects had normal hearing based on past hearing screens. Participants either had normal vision or wore corrective lenses. Participants with a history of seizure disorder, significant hearing or visual impairment, a brain abnormality conclusive from imaging studies or an identified genetic disorder were excluded. All participants were high-functioning children with ASD with full scale IQ > 80 assessed using the Wechsler Intelligence Scale for Children, Fourth Edition (WISC-IV; [24]), the Stanford-Binet Intelligence Test [25], or the Wechsler Abbreviated Scale of Intelligence (WASI, [26]).

Typically developing (TD) children were recruited through advertisements in the local media. All TD participants were free of neurological or significant medical disorders, had normal hearing and vision, and were free of psychiatric, learning, or developmental disorders based on self- and parent reports. Subjects were screened for history of psychiatric or neurological diagnosis using the Structured Clinical Interview for DSM-IV Non-Patient Edition (SCID-NP, [27]). Participants within the control and autism groups were attempted to be matched for age, gender, and the socioeconomic status of their family.

The mean age of 32 participants enrolled in the ASD group was 13.09 ± 2.41 years (range 9–18 years, 29 males, 3 females), while the mean age of the TD group (N = 24) was 13.91 ± 2.91 years (9–20 years, 20 males, 4 females). The age difference between groups was not significant ($p = 0.38$, n.s.). All children with autism were high functioning individuals. Mean Full Scale IQ score for children with autism was 90.1 ± 14.9 and were collected from their most recent IQ evaluation records. Most of children in TD group did not have their IQ records available and their mean IQ data was not possible to report. Socioeconomic status of ASD and control groups was compared based on parent education and annual household income. The approximate household incomes did not reveal any statistically significant group differences. Participants in both groups had similar parent education levels.

The study complied with all relevant national regulations and institutional policies and was approved by the local Institutional Review Board (IRB). Participating subjects and their parents (or legal guardians) were provided with full information about the study including the purpose, requirements, responsibilities, reimbursement, risks, benefits, alternatives, and role of the local IRB. The consent and assent forms approved by the IRB were reviewed and explained to all subjects who expressed interest to participate. All questions were answered before consent signature was requested. If the individual agreed to participate, both she/he and parent/guardian signed and dated the consent form and/or assent form and received a copy countersigned by the investigator who obtained consent.

2.2. ERP Data Acquisition, and Signal Processing

Electroencephalographic (EEG) data was acquired with a 128 channel Electrical Geodesics Inc. (EGI) system (v. 200) consisting of Geodesic Sensor Net electrodes, Net Amps and Net Station software (Electrical Geodesics Inc., Eugene, OR, USA) running on a Macintosh G4 computer. EEG data were sampled at 500 Hz and 0.1–200 Hz analog filtered. Impedances were kept under 40 KΩ. According to the Technical Manual of EGI [28] this Net Sensor electrode impedance level is sufficient for quality recording of EEG with this system. The Geodesic Sensor Net is a lightweight elastic thread structure containing Ag/AgCl electrodes housed in a synthetic sponge on a pedestal. The sponges were soaked in a KCl solution to render them conductive. EEG data was recorded continuously. EEG channel with high impedance or with visually detectable artifacts (e.g., channel drift, gross movement, etc.) were marked as bad using the Net Station event marker tools in "on-line" mode for further removal in the "off-line" mode using the Net Station Waveform Tools (NSWT).

Stimulus-locked EEG recordings were segmented off-line into 1000 ms epochs spanning 200 ms pre-stimulus to 800 ms post-stimulus around the critical stimulus events: e.g., in an oddball task: (1) rare target, (2) rare non-target distracter (novel), (3) frequent non-target (standard). Data sets were digitally screened for artifacts (eye blinks, movements), and contaminated trials were removed using artifact rejection tools. The Net Station Waveform Tools' Artifact Detection module in "off-line" mode marks EEG channel bad if fast average amplitude exceeds 200 μV, differential average amplitude exceeds 100 μV or if channel has zero variance. Segments were marked bad if they contained more than 10 bad channels, or if eye blink or eye movement were detected (>70 μV). After detection of bad channels, the NSWT's "bad channel replacement" function was used for the replacement of data in bad channels with data interpolated from the remaining good channels (or segments) using spherical splines [29–32]. Response-locked ERPs were segmented off-line into 1000 ms epochs spanning 500 ms pre-error motor response to 500 ms post around the critical response event—commission error. Remaining data were digitally filtered using 60 Hz Notch and 0.3–20 Hz bandpass filters and then segmented by condition and averaged to create ERPs. Averaged ERP data were baseline corrected and re-referenced into an average reference frame. All stimuli and behavioral response collection was controlled by a PC computer running E-prime software (Psychology Software Tools Inc., Sharpsburg, PA, USA). Visual stimuli were presented on a 15 inch display. Manual responses were collected with a 5-button keypad (Serial Box, Psychology Software Tools Inc., Sharpsburg, PA, USA).

2.3. Three Stimuli Visual Oddball Test with Novel Distracters

This test represented a modification of traditional visual three-stimulus oddball task. Stimuli letters "X", "O", and novel distracters ("v", "^", ">" and "<" signs) were presented on the screen after fixation mark "+". One of the stimuli ("O") was presented on 50% of the trials (frequent standard); the novel stimuli stimulus (e.g., ">") was presented on 25% of the trials (rare distracter), whereas the third ("X") was presented on the remaining 25% of the trials and represented the target. Subjects were instructed to press a key when they see the target letter on the screen. Each stimulus was presented for 250 ms, with a variable (1000–1100 ms) inter-trial interval. There were 480 trials in total, with a break every 240 trials. The complete sequence took around 20 min.

2.4. Behavioral Measures

Behavioral response measures were mean reaction time (RT in ms) and response accuracy (percent of correct hits). Number and percent of commission and omission errors along with total number of errors was calculated for each participant. Post-error RT was calculated as RT to the first correct response after committed error (either omission of commission error). Difference between post-error RT and preceding correct RT to target was used to calculate normative post-error RT slowing values [5,8,33]. Analysis of distribution of RT in correct and error trials was conducted using sigmoid curve methodology from Opris et al. [34].

2.5. Event-Related Potentials (ERP)

2.5.1. Stimulus-Locked ERPs

ERP dependent measures were: adaptive mean amplitude and latency of ERP peak (e.g., P3a, P3b) within a temporal window across a region-of-interest (ROI) channel group. A list of ERP dependent variables included stimulus-averaged amplitude and latency of the frontal ERP components: N1 (80–180 ms post-stimulus), N2 (200–320 ms), and P3a (300–520 ms); and the posterior (parietal ROIs) ERP components P1 (80–180 ms), N2 (N2b, 180–320 ms) and P3b (320–560 ms). The frontal ROIs for N1, N2 and P3a components included following EGI channels: left ROI—EGI channel 12 (between FC1 and FCz), F1, F3 and FC1; midline ROI—FCz, Fz; right ROI—EGI channel 5 (between F2 and FCz),

F2, F4 and FC2. The parietal ROI for P1, N2 and P3b components included following EGI channels: left ROI—P1, P3, P7, PO3; midline—Pz; right ROI—EGI channel P2, P4, P8, PO8.

2.5.2. Response-Locked Dependent Variables (ERN/Pe)

Response locked dependent variables in this study were adaptive mean amplitude and latency of the Error-related Negativity (ERN peaking within 40–150 ms post-error) and Error-related Positivity (Pe, peaking within 100–300 ms post-error). The ROI for both ERN and Pe components included FCz, sites between FCz and FC3-C1, and between FCz and FC2-C2).

2.6. Statistical Data Analysis

Statistical analyses were performed on the subject-averaged behavioral (RT, error rate) and ERP data with the subject averages being the observations. The primary analysis model was the repeated measures ANOVA, with dependent variables being reaction time (RT), accuracy, commission and omission error rate, post-error RT slowing index, RT in correct and commission error trials, response locked ERN and Pe, and all the specific stimulus-locked ERP components' (N1, P1, N2, P3) amplitudes and latencies at the selected frontal and parietal ROIs. The data of each ERP dependent variable for each relevant ROI was analyzed using ANOVA with the following factors (all within-participants): *Stimulus* (target, novel, standard) and *Hemisphere* (left, right). The between subject factor was *Group* (ASD, CNT). Histograms with normal distribution curves along with skewness and kurtosis data were obtained for each dependent variable to determine normality of distribution and appropriateness of data for ANOVA test. In all ANOVAs, Greenhouse-Geisser corrected *p*-values were employed where appropriate. Since ASD group differences from TD group were the most important aspects of the study, the focus of analysis was on main effects of *Group* (ASD vs. TD) and their interactions with other variables. For the estimation of the effect size [35] we used Partial Eta Squared (η_p^2) measure. Statistical analysis was performed using SPSS v.22 and Sigma Stat 3.1 packages.

3. Results

3.1. Behavioral Responses

Reaction time to targets in autism group was not different from TD group (473.2 ± 93.5 vs. 465.6 ± 98.4, $F_{1,54}$ = 0.08, *p* = 0.78, n.s.), but the difference in commission error rate was significant (5.57 ± 9.32 percent in ASD vs. 1.12 ± 1.20 percent in TD, $F_{1,54}$ = 4.48, *p* = 0.040), being higher in the ASD group. The difference between mean RT in correct trials and post-error trials (i.e., post-error RT -minus-correct trial RT) was negative in autism but positive in controls, and this between-group difference was significant (−15.7 ± 50.1 ms in ASD vs. 19.4 ± 37.9 ms in TD, $F_{1,54}$ = 6.00, *p* = 0.019, see Figure 1). Analysis of distribution of RT in correct and error trials using sigmoid curve method [34] did not show group differences, however in the ASD group distribution of RT in the error trials tended to be in the faster bin range (moda ~200 ms in error vs. ~400 ms in correct trials, see Figure 2.)

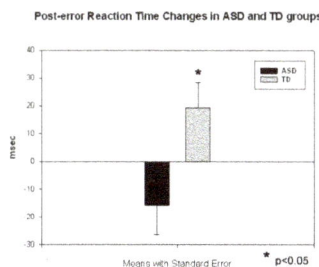

Figure 1. Post-error reaction time (RT) changes (calculated as a difference between the first post-error RT minus mean preceding RT) in ASD and TD children. Typical children show normative post-error RT slowing, conversely ASD children respond faster after having committed an error.

Correct vs Error RT for ASD

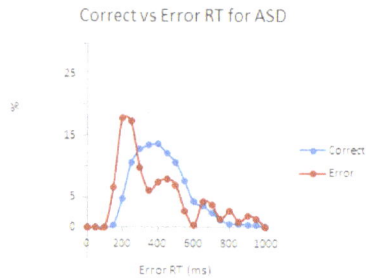

Figure 2. Distribution of RT in correct and error trials in children with ASD. Error trials had higher percentage in the faster bins of the histogram with moda around 200 ms vs. 400 ms in correct trials.

3.2. Motor Response-Locked Fronto-Central ERPs

ERN and Pe. One subject from the ASD group and 4 subjects from the TD group had no commission errors, therefore ERN/Pe datasets were comprised of 31 autistic children and 20 TD controls. Amplitude of the midline fronto-central ERN was significantly less negative in the ASD group as compared to the TD group (-2.69 ± 6.17 µV vs. -6.36 ± 4.43 µV, $F_{1,49} = 4.94$, $p = 0.031$). Latency of the ERN in the ASD group was delayed (107.5 ± 36.1 ms vs. 80.1 ± 20.7 ms, $F_{1,49} = 9.46$, $p = 0.003$). The only group difference in Pe measure was detected for the latency, though this difference barely reached minimal significance level and was featured by the tendency for prolonged latency in the ASD group (206.4 ± 44.9 ms vs. 183.3 ± 33.3 ms, $F_{1,49} = 4.05$, $p = 0.05$).

3.3. Stimulus-Locked Event-Related Potentials

3.3.1. Frontal ERPs

N1. *Stimulus* (target, standard, novel) factor had large main effect on N1 amplitude in both hemispheres ($F_{2,53} = 7.26$, $p = 0.002$, $\eta_p^2 = 0.22$). Amplitude of the frontal N1 to targets was bilaterally more negative in ASD group as compared to TD group (-3.68 ± 2.07 µV vs. -1.79 ± 2.29 µV, $F_{1,54} = 10.01$, $p = 0.003$). The ASD group showed as well more negative N100 amplitude to standards (-4.22 ± 1.71 µV vs. -2.39 ± 2.15 µV, $F_{1,54} = 12.05$, $p = 0.001$) and novels (-4.74 ± 2.02 µV vs. -2.61 ± 2.09 µV, $F_{1,54} = 13.87$, $p < 0.001$). Figure 3 illustrates group differences in response to target and novel stimuli. There were no interactions to report. Latency of the frontal N1 did not show any statistical group differences.

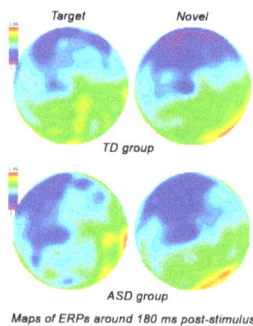

Maps of ERPs around 180 ms post-stimulus

Figure 3. Grandaverage qEEG map of ERP to target and novel stimuli around 180 ms post-stimulus in ASD and TD children. Children with ASD showed comparable high negativity (N1 ERP component) to both targets and novels.

N2a. Most pronounced group differences resulted when target and novel stimuli effects were compared. Stimulus type had medium main effect on amplitude of the bifrontal N2 ($F_{2,53} = 4.84$, $p = 0.032$, $\eta_p^2 = 0.09$). The ASD group in response to targets showed more negative amplitude bilaterally (-3.44 ± 3.45 µV vs. -1.46 ± 2.47 µV, F = 5.22, $p = 0.027$). There was significant *Stimulus* (target, novel) X *Group* (ASD, TD) interaction ($F_{1,54} = 4.82$, $p = 0.033$, $\eta_p^2 = 0.08$), that can be described as comparable N200 amplitude to targets and novels in the ASD group, while in the TD group amplitude was larger for novel stimuli. Latency of N200 component was significantly longer in the ASD group across all 3 categories of stimulation without any hemispheric differences (targets, $F_{1,54} = 9.33$, $p = 0.004$; standards, $F_{1,54} = 6.84$, $p = 0.012$; novels, $F_{1,54} = 6.96$, $p = 0.011$).

P3a (Novelty P3). Rare *Stimulus* (target, novel) type had moderate main effect on P3a amplitude across both hemispheres ($F_{2,53} = 4.22$, $p = 0.041$, $\eta_p^2 = 0.08$). Group differences were found only for novel distracters, in particular, the ASD group had bilaterally higher amplitude as compared to the TD group (5.03 ± 3.33 µV vs. 2.88 ± 3.96 µV, $F_{1,54} = 4.44$, $p = 0.04$). Children with ASD as compared to the TD children had as well higher P3a amplitude to standards ($F_{1,54} = 7.56$, $p = 0.008$), but amplitude of P3a did not show any group differences in response to targets. Group differences for P3a latency were found only in the left hemisphere for all three categories of stimuli, and all of them featured a more prolonged latency in the ASD group (left frontal ROI- targets, $F_{1,54} = 4.88$, $p = 0.031$; standards, $F_{1,54} = 5.07$, $p = 0.028$; novels, $F_{1,54} = 4.73$, $p = 0.034$, see Figures 4 and 5). *Hemisphere* (left, right) X emphGroup (ASD, TD) interaction was significant, and can be described as more prolonged latency of the P3a component at the left hemisphere in the ASD group ($F_{1,54} = 6.73$, $p = 0.012$, $\eta_p^2 = 0.11$).

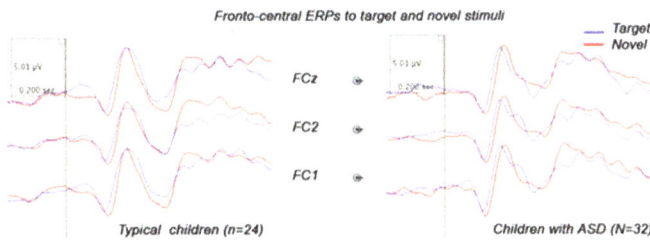

Figure 4. Screenshot of fronto-central ERPs to target and novel stimuli in ASD and TD children. The ASD children showed more negative N1, prolonger N2a and augmented P3a in response to unattended novel distracters.

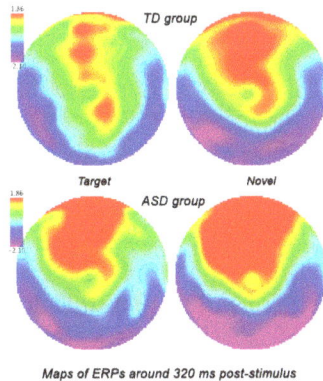

Figure 5. Grandaverage qEEG map of ERP to target and novel stimuli around 320 ms post-stimulus in ASD and TD children. Children with ASD showed higher positivity (P3a ERP component) to both targets and novels at the frontal and froto-central topographies.

3.3.2. Parietal and Parieto-Occipital ERPs

P1. *Stimulus* type (target, standard, novel) had marginal bilateral main effect on amplitude of the posterior P100 component ($F_{2,53}$ = 3.20, p = 0.049, η_p^2 = 0.05). Group differences were expressed in a higher amplitude in the ASD group to standards (3.57 ± 2.08 µV vs. 2.33 ± 2.41 µV, $F_{1,54}$ = 4.15, p = 0.046) and to novels (4.24 ± 2.36 vs. 2.55 ± 2.26, $F_{1,54}$ = 7.34, p = 0.009, see Figure 6). Comparison of P1 amplitude to target and novel stimuli showed moderate *Stimulus* X *Group* interaction ($F_{1,54}$ = 4.36, p = 0.042, η_p^2 = 0.08) where the ASD group showed higher response to novels as compared to the TD group. There were found no latency group differences for this component.

Left parietal ERP to novels in TD and ASD children

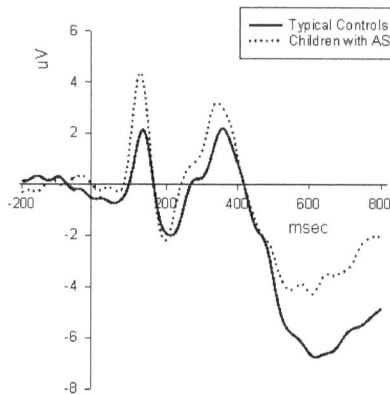

Figure 6. Left parietal ERPs to novels in ASD and TD children. The ASD group showed higher amplitude of P1 and P3b ERP components in response to novel distracters.

N2b. Amplitude of the parietal N2b component did not show any group differences. On the other hand, latency of N2b was globally delayed in the ASD group to all stimuli, and this effect was more pronounced on the left ROI (targets, 257.1 ± 34.1 ms in ASD vs. 222.5 ± 41.0 ms in TD, $F_{1,54}$ = 10.78, p = 0.002; standards, 253.7 ± 40.2 ms vs. 211.1 ± 59.9 ms, $F_{1,54}$ = 9.38, p = 0.004; novels, 248.1 ± 35.6 ms vs. 216.5 ± 53.5 ms, $F_{1,54}$ = 6.45, p = 0.014). *Stimulus* (target, novel) X *Group* interaction was significant ($F_{1,54}$ = 6.08, p = 0.017, η_p^2 = 0.11) and this effect was featured by a prolonged latency to targets in the ASD group.

P3b. *Stimulus* type had large main effect on P3b amplitude in both hemispheres ($F_{2,53}$ = 43.75, p < 0.001, η_p^2 = 0.62). Analysis of the parietal P3b amplitude yielded group differences only in response to novel stimulus (7.24 ± 3.02 µV in ASD vs. 4.93 ± 4.32 µV in TD, $F_{1,54}$ = 5.63, p = 0.021). Stimulus factor also had strong main effect on the latency of P3b peak ($F_{2,53}$ = 8.76, p < 0.001; η_p^2 = 0.24). However, we didn't find any P3b latency group differences or any significant interactions.

4. Discussion

The reaction time findings in this study indicate that children with ASD had a less accurate behavioral performance. Committed motor response errors were not followed by a normative post-error slowing of reaction time. Error-related negativity was less pronounced and prolonged in the ASD group, though error-relative positivity only tended to be delayed. Both early (P1, N1) and late (P3a, P3b) ERP components in response to novel distracters were enhanced and showed larger amplitude in the ASD group when compared to the TD group. The N2 component showed delayed latency to all categories of stimuli at the frontal and parietal regions of interest. Results partially replicate our earlier ERP findings in autism. Our prior studies showed similar group differences in reaction time tasks using two different types of visual three-stimulus oddball tests, including one with illusory Kanizsa figures as the stimuli [3–6,8,9].

In numerous research studies and reviews autistic children have been found to differ from typical children mainly with respect to the P3 in regular oddball tasks (reviewed in Cui et al. [36]). Kemner et al. [37–39] reported abnormally small occipital and reduced central P3 in response to visual stimuli. At the same time, these authors also reported that the parietal N2 and P3b was larger in autistic children and interpreted their results in terms of differences in attentional resource allocation, as the parietal P3b is more sensitive to such task manipulations as stimulus relevance and probability [40] and less dependent on modality as compared to the occipital P3 [39]. In general, studies using a simple visual target detection, as compared to cross-modal (e.g., audio-visual integration) tasks have found no significant differences in the P3b to targets in children with autism compared to typical controls, while abnormalities were present in dissociations of frontal (delayed) and posterior (relatively intact) P3 in visual attention tasks [41]. It should be noted, that in most ERP studies using oddball tasks, the focus was on ERP responses to target stimuli, and less attention was paid to atypically large ERP magnitude to non-target stimuli (i.e., standards and/or novel distracters). Our results in this study, as well as in our prior research studies [3–9] emphasize the need to perform more detailed analysis of the specifics of excessive reactivity to task-irrelevant items in oddball tests in children with ASD.

The finding of increased amplitude of the frontal P3a to novels in children with autism is of special interest as this ERP measure is directly related to orienting to novelty. The frontal novelty P3a is less explored in autism and results are not consistent [41–43]. It was reported that the frontal P300 which reflects attention orienting was delayed or missing in subjects with autism and this finding was interpreted by the authors as a disruption of both parieto-frontal and cerebello-frontal networks critical for efficient cross-modal integration. Kemner et al. [37] reported that the visual N200 to novel distracters is larger when a person with autism is performing a task even when these novel stimuli are not relevant to the task in question. Abnormalities in central sensory processing both in auditory and visual modalities have been described by different authors in autism [37,39,43–45]. However, most of these studies analyzed and reported findings relating P3b to targets [42,46,47] and only a few P3a to standards or novels [41,43]. Despite the large number of studies published on ERPs in autism, there are not many reports about ERP components similar to those analyzed in this study. Most of the studies outline hyperactivation as well as an abnormal pattern of primary perceptual processes (e.g., low selectivity), abnormal top-down attentional control (e.g., orienting to novelty) and irregular information integration processes [48,49]. In control subjects the frontal P3a occurs earlier and commonly precedes parietal P3b, but in autistic subjects the P3a and P3b components were found to peak almost at the same time over the frontal and parietal sites in a spatial visual attention task [41]. In our study, novel stimuli elicited a delayed P3a component in the autism group. The latency of this component usually is associated with speed of attentional orienting to stimulus and reflects prefrontal working memory processes. Even less explored are early ERPs potentials in oddball tasks in ASD. The role of exogenous components and contribution to abnormalities of behavior in autism deserves further investigation.

In our prior ERP studies [3,5] on novelty processing in ASD, we reported that children with ASD showed significantly higher amplitudes and longer latencies of early frontal ERPs and delayed latency of P3a to novel distractor stimuli. The current study replicates those results mostly in terms of larger amplitudes of early potentials (P1, N1) and augmented magnitude of late potentials (P3a, P3b) to novel distracters. Our results suggest low selectivity in modality specific pre-processing of stimulus. Higher amplitude of the early frontal negativity and parietal positivity in the autism group with minimal differentiation of response magnitude to either target or non-target stimuli is an interesting finding that replicates our previous reports [6–9] where different types of visual oddball tasks were used. The visual N1 is considered to be an index of stimulus discrimination [50]. The visual N1 generally is augmented during pre-attentional stimulus processing [51], and is larger towards task-relevant target stimuli rather than unattended stimuli [52]. The ASD group shows clearly augmented and delayed frontal P3a that might result in an impaired early differentiation of target and non-target items (e.g., on N1 stage) and more effortful compensatory strategies involved for successful target identification, and following correct motor response selection. In general, the autistic group showed prolonged latencies to standard and rare non-target cues in visual oddball task. These results suggest that individuals with autism probably over-process information needed for the successful differentiation of target and distractor stimuli.

Cortical activity during different stages of visual information processing can be detected with ERPs, as they represent stimulus-driven corticoelectric field potentials. The early ERP components are a series of potentials that are recorded at the scalp following sensory stimulation that usually occur between 40 and 200 ms post-stimulation. These components are also characterized by being exogenous in nature (i.e., they are predictably generated by delivering sensory stimulation without a need for the subject to perform any mental operations). This characteristic differentiates the early latency ERPs from endogenous ERPs (e.g., P3b), which require the performance of a cognitive task such as a novelty task used in our study. Dysfunctional selective filtering of the stimuli may occur at the levels of P1 and N1 response related to the sensory gating stage. Sensory gating is based on selective attention concepts as attention towards one stimulus requires automatic concurrent inhibition of attention towards another one. Habituation to irrelevant sensory input is an important function for the information processing by the brain, because the failure might be associated with mental disturbances. The relationship between sensory gating and oversensitivity of individuals with autism to sensory stimulation other than auditory stimulation remains understudied and the clinical correlates of visual P100 and N100 ERP components in autism are yet to be examined in-depth.

A number of potentially interesting correlations were found between P1 and N1 sensory gating measures and P3 variables in schizophrenia [53]. Among these correlations, the positive interdependence between the N1 and P3 (P3b) latency is of interest as it is relevant to the findings of our previous [3,5,8] and current study. The study of Boutros et al. [53] suggests that decreased gating at the N1 phase of information processing negatively impacts the speed of information processing as measured by P3b latency. The further observation that this correlation was stronger in schizophrenia patients [53] emphasizes the possible deleterious effects of abnormal P1 and N1 level gating on information processing. This finding raises the possibility that the early sensory gating deficit may also impact the resource allocation capacity of the cortex as measured by P3b amplitude.

Belmonte and Yurgelin-Todd [48,49] suggested that perceptual filtering of incoming stimulation in autism can be considered as occurring in an all-or-none manner with low specificity for the task relevance of the stimulus. This notion assumes that filtering of perceptual items may primarily be driven by the control of general arousal level rather than the activation of only modality-specific cortical areas. According to the latter authors the attention of individuals with autism seems to be founded more on the coarse control of general arousal than on selective activation of specific perceptual systems. It is reasonable to suggest that an active inhibition of irrelevant distracters is not properly functioning and allows both task-relevant and task-irrelevant stimuli to pass through earlier filtering processes creating the overload on the later stages of stimuli processing in the context of task. It is unsurprising in this context that an increased ratio of excitation/inhibition in key neural systems and high "cortical noise" has been considered as a core abnormality of autism [54,55].

In this study, and similar to our studies on error monitoring in autism [7,8], we found that the ERN component of the response-locked ERP was substantially decreased in children with autism. In particular, the amplitude of ERN was less negative and the latency of ERN was prolonged in the ASD group as compared to the TD children. The ERN is an electroencephalographic measure associated with the commission of errors, thought to be independent of conscious perception [33], while the Pe is thought to reflect the motivational or emotional significance of the error or, in other words, the conscious evaluation of the error [56]. It cannot be ruled out that ERN impairments are influenced by deficits in earlier perceptual processes, or attentional and working memory processes in children with autism, that might be reflected in altered stimulus-locked early and late ERPs.

It has been suggested [57,58] that both the response-locked ERN and the stimulus-locked frontal N2 might reflect similar processes (i.e., response conflict detection and monitoring) and have similar neural correlates (i.e., dipole in the ACC, see also van Veen and Carter [59]). On the behavioral level, we found no group differences in RT, and only group differences between the percentages of commission (and not omission) error in the visual novelty oddball. After an error, ASD patients did not show accuracy improvement through post-error RT slowing as typical controls did. This finding replicates our previous reports [5–8]. Normally, performance on these trials is improved as a result of a change in speed–accuracy strategy which reflects executive control functioning [60]. The atypical post-error performance of ASD children (i.e., speeding instead of normative slowing) suggests the presence of an executive control deficiency. The impairment of adaptive error-correction behavior may have important consequences in daily life as optimal error-correction is necessary for adequate behavioral responses.

As demonstrated in previous studies [61], the posterior medial frontal cortex, and more specifically the rostral ACC division, is the main brain area responsible for error processing, suggesting that ASD patients have reduced posterior medial frontal cortex functioning. This area is involved when there is a need for adjustments to achieve goals [61]. The findings pointing that children with ASD have an impaired ability to improve their response accuracy by slowing down the response speed on post-error trials agrees with this notion. However, it is necessary to take into account that observed significant group differences between ASD and typical controls are manifested not only in the behavioral performance measures on reaction time tasks and associated response monitoring indices such as ERN and Pe. Group differences were also noted in terms of amplitude and latency characteristics of early ERP components preceding motor response selection (frontal and parietal P1, N1, N2) and those reflecting context update and closure (e.g., P3b) in visual oddball task [5] and various auditory tasks [44]. The sum of the group differences across these behavioral and stimulus- and response-averaged ERP indices of the ASD patients' performance is that it reflects global deficits in attentional processes, more specifically deficits in effective differentiation of target and novel distracter stimuli. This latter interpretation is supported by the significant differences between the ASD patients and typically developing controls in terms of both the stimulus-locked and response-locked ERP amplitudes and latencies.

Post-error adaptive correction of responses might be explained by some recent neurobiological findings. There are reports about an excessive preservation of short-distance connections (i.e., local over-connectivity) and relatively poor long-distance connections (i.e., distant under-connectivity) in the neocortex of individuals with autism [62–65]. These cortical connectivity abnormalities may explain why persons with autism tend to focus on details rather than perceiving the whole Gestalt. This over-focusing on details may imply an excessively laborious and ineffective way of handling each trial in the cognitive test, and lower availability of resources after an error when effort is needed to react appropriately. This may result in insufficient activation of the ACC [66], and thus error detection and post-error reaction may be hampered [67,68]. Structural and functional deficiencies of the ACC may contribute to the atypical development of joint attention and social cognition in autism [69]. Such interpretation of the results of the ERN/Pe deficits found in several studies [6,7,66,70] is consistent with many aspects of theory and research suggesting that ACC-mediated response monitoring may contribute to social-emotional and social-cognitive development in autism [69]. However, while emphasizing the possible role of ACC-related self-monitoring deficits in autism, Mundy [69] also noted that according to Devinsky and Luciano [71] this ACC impairment related behavioral deficits emerge only when they are combined with disturbances in other related functional neural networks, e.g., dorsolateral prefrontal cortex (DLPFC). Another factor that might affect variability of ERN is related to developmental changes, as performance monitoring in children and adolescents endures changes due to maturation processes [72], while our sample included both young children (9–12 years old range) and adolescents (i.e., teenagers in 13–18 years old range) and the wide range of participants in our study should be considered as a certain limitation.

Abnormalities of early exogenous ERPs such as N1 and P1 components can negatively affect endogenous potentials (e.g., N2 and P3) as well as response-locked potentials (ERN and Pe). The deficits in identification of distinct categories of stimuli at the early sensory stage result in a need to delegate task of differentiation of target stimulus from irrelevant one to the later, higher level of information processing stage. It is possible to suggest that individuals with ASD may engage some compensatory strategies necessary for successful target detection.

5. Conclusions

The results of the study indicate an atypical manner of processing distracters and orienting attention to novelty in children with autism. The findings are in-keeping with our prior studies using different tasks with visual and auditory stimuli. Augmented early potentials and a delayed frontal P3a and parietal P3b to novel stimuli suggest low selectivity in pre-processing of distracters resulting in excessive information processing at later stages. This may indicate a reduction in the discriminative ability of the ASD group. These results may reflect a locally over-connected network where sensory inputs evoke abnormally large ERPs for unattended stimuli with signs of a reduction in selectivity. This may incur a high load at the later stages of perceptual and cognitive processing and response selection when novel distracter stimuli are differentiated from targets.

Acknowledgments: Funding for this study was provided by the NIH grant R01 MH086784 to Manuel F. Casanova.

Author Contributions: All authors contributed substantially to the work reported in this paper. Manuel Casanova and Estate Sokhadze conceived and designed the experiment and data analysis strategy, participated in oddball test data collection, statistical analysis and interpretation of results. Estate Sokhadze administered visual oddball test to patients and collected raw data. Eva Lamina and Emily Casanova participated in ERP data collection, preprocessing and statistical analysis and also contributed in preparation of figures; Irma Khachidze participated in reaction time and accuracy data processing and statistical analysis; Ioan Opris analyzed reaction time data and assisted in preparation of methodological part related to behavioral data handling, analysis and interpretation; Manuel Casanova and Desmond Kelly participated in patients clinical records analysis and in interpretation of data and discussion, they also reviewed the manuscript and edited the final version.

Conflicts of Interest: The authors declare no conflict of interest.

References

1. *Diagnostic and Statistical Manual of Mental Disorders (DSM-V)*, 5th ed.; American Psychiatric Publishing, Inc.: Washington, DC, USA, 2013; ISBN 089042554X.
2. Gomes, E.; Pedroso, F.S.; Wagner, M.B. Auditory hypersensitivity in the autistic spectrum disorder. *Pro Fono* **2008**, *20*, 279–284. [CrossRef] [PubMed]
3. Baruth, J.M.; Casanova, M.F.; Sears, L.; Sokhadze, E. Early-stage visual processing abnormalities in autism spectrum disorder (ASD). *Transl. Neurosci.* **2010**, *1*, 177–187. [CrossRef] [PubMed]
4. Casanova, M.F.; Baruth, J.M.; El-Baz, A.; Tasman, A.; Sears, L.; Sokhadze, E. Repetitive transcranial magnetic stimulation (rTMS) modulates event-related potential (ERP) indices of attention in autism. *Transl. Neurosci.* **2012**, *3*, 170–180. [CrossRef] [PubMed]
5. Sokhadze, E.; Baruth, J.; Tasman, A.; Sears, L.; Mathai, G.; El-Baz, A.; Casanova, M.F. Event-related potential study of novelty processing abnormalities in autism. *Appl. Psychophysiol. Biofeedback* **2009**, *34*, 37–51. [CrossRef] [PubMed]
6. Sokhadze, E.; Baruth, J.; Tasman, A.; Mansoor, M.; Ramaswamy, R.; Sears, L.; Mathai, G.; El-Baz, A.; Casanova, M.F. Low-frequency repetitive transcranial magnetic stimulation (rTMS) affects event-related potential measures of novelty processing in autism. *Appl. Psychophysiol. Biofeedback* **2010**, *35*, 147–161. [CrossRef] [PubMed]
7. Sokhadze, E.; Baruth, J.; El-Baz, A.; Horrell, T.; Sokhadze, G.; Carroll, T.; Tasman, A.; Sears, L.; Casanova, M. Impaired error monitoring and correction function in autism. *J. Neurother.* **2010**, *14*, 79–95. [CrossRef] [PubMed]

8. Sokhadze, E.M.; Baruth, J.M.; Sears, L.; Sokhadze, G.E.; El-Baz, A.S.; Williams, E.; Klapheke, R.; Casanova, M.F. Event related potentials study of attention regulation during illusory figure categorization task in ADHD, autism spectrum disorders, and typical children. *J. Neurother.* **2012**, *16*, 12–31. [CrossRef] [PubMed]

9. Sokhadze, E.M.; Baruth, J.; Tasman, A.; Casanova, M.F. Event-related potential studies of cognitive processing abnormalities in autism. In *Imaging the Brain in Autism*; Casanova, M.F., El-Baz, A.S., Suri, J.S., Eds.; Springer: New York, NY, USA, 2013; pp. 61–86, ISBN 978-1-4614-6843-1.

10. Casanova, M.F.; Sokhadze, E.; Opris, I.; Wang, Y.; Li, X. Autism spectrum disorders: Linking neuropathological findings to treatment with transcranial magnetic stimulation. *Acta Pediatr.* **2015**, *104*, 346–355. [CrossRef] [PubMed]

11. Sokhadze, E.M.; Casanova, M.F.; Casanova, E.; Kelly, D.P.; Khachidze, I.; Wang, Y.; Li, X. Applications of ERPs in autism research and as functional outcomes of neuromodulation treatment. In *Event-Related Potential (ERP): Methods, Outcomes and Research Insights*; Harris, S.R., Ed.; NOVA Science Publishers: Hauppauge, NY, USA, 2017; pp. 27–88, ISBN 1536108057.

12. Duncan, C.C.; Barry, R.J.; Connolly, J.F.; Fischer, C.; Michie, P.T.; Näätänen, R.; Polich, J.; Reinvang, I.; van Petten, C. Event-related potentials in clinical research: Guidelines for eliciting, recording, and quantifying mismatch negativity, P300, and N400. *Clin. Neurophysiol.* **2009**, *120*, 1883–1908. [CrossRef] [PubMed]

13. Picton, T.W. The P300 wave of the human event-related potential. *J. Clin. Neurophysiol.* **1992**, *9*, 456–479. [CrossRef] [PubMed]

14. Polich, J. Updating P300: An integrative theory of P3a and P3b. *Clin. Neurophysiol.* **2007**, *118*, 2128–2148. [CrossRef] [PubMed]

15. Pritchard, W. Psychophysiology of P300. *Psychol. Bull.* **1981**, *89*, 506–540. [CrossRef] [PubMed]

16. Coles, M.G.H.; Rugg, M.D. Event-related brain potentials: An introduction. In *Electrophysiology of Mind. Event-Related Brain Potentials and Cognition*; Rugg, M.D., Coles, M.G.H., Eds.; Oxford University Press: Oxford, UK, 1995; pp. 1–26, ISBN 0198524161.

17. Herrmann, C.S.; Knight, R.T. Mechanisms of human attention: Event related potentials and oscillations. *Neurosci. Biobehav. Rev.* **2001**, *25*, 465–476. [CrossRef]

18. Patel, S.H.; Azzam, P.N. Characterization of N200 and P300: Selected studies of the event-related potential. *Int. J. Med. Sci.* **2005**, *2*, 147–154. [CrossRef] [PubMed]

19. Hoffman, J.E. Event-related potentials and controlled processes. In *Event-Related Brain Potentials: Basic Issues and Applications*; Rohrbaugh, J.W., Parasuraman, R., Johnson, R., Jr., Eds.; Oxford University Press: New York, NY, USA, 1990; pp. 145–157, ISBN 0195048911.

20. Katayama, J.; Polich, J. Stimulus context determines P3a and P3b. *Psychophysiol.* **1998**, *35*, 23–33. [CrossRef]

21. Polich, J. Theoretical overview of P3a and P3b. In *Detection of Change: Event-related Potential and fMRI Findings*; Polich, J., Ed.; Kluwer Academic Press: Boston, MA, USA, 2003; pp. 83–98, ISBN 978-1-4613-5008-8.

22. American Psychiatric Association. *Diagnostic and Statistical Manual of Mental Disorders (DSM-IV-TR)*, 4th ed.; Text Revised; American Psychiatric Publishing, Inc.: Washington, DC, USA, 2000; ISBN 0890420629.

23. Le Couteur, A.; Lord, C.; Rutter, M. *The Autism Diagnostic Interview—Revised (ADI-R)*; Western Psychological Services: Los Angeles, CA, USA, 2003.

24. Wechsler, D. *Wechsler Intelligence Scale for Children—Fourth Edition (WISC-IV)*; Harcourt Assessment, Inc.: San Antonio, TX, USA, 2003.

25. Roid, G.H. *Stanford-Binet Intelligence Scales Technical Manual*, 5th ed.; Riverside Publishing: Itasca, IL, USA, 2003.

26. Wechsler, D. *Wechsler Abbreviated Scale for Intelligence*; Psychological Corporation: San Antonio, TX, USA, 1999.

27. First, M.B.; Spitzer, R.L.; Gibbon, M.; Williams, J.B.W. *Structured Clinical Interview for DSM-IV-TR Axis I Disorders—Non-Patient Edition (SCID-NP)*; New York State Psychiatric Institute: New York, NY, USA, 2002.

28. *Net Station Acquisition Technical Manual*; Electrical Geodesics, Inc.: Eugene, OR, USA, 2003.

29. Fletcher, E.M.; Kussmaul, C.L.; Mangun, G.R. Estimation of interpolation errors in scalp topographic mapping. *Electroctoencephalogr. Clin. Neurophysiol.* **1996**, *98*, 422–434. [CrossRef]

30. Luu, P.; Tucker, D.M.; Englander, R.; Lockfeld, A.; Lutsep, H.; Oken, B. Localizing acute stroke-related EEC changes: assessing the effects of spatial undersampling. *J. Clin. Neurophysiol.* **2001**, *18*, 302–317. [CrossRef] [PubMed]

31. Perrin, E.; Pernier, J.; Bertrand, O.; Giard, M.; Echallier, J.F. Mapping of scalp potentials by surface spline interpolation. *Electroencephalogr. Clin. Neurophysiol.* **1987**, *66*, 75–81. [CrossRef]

32. Srinivasan, R.; Tucker, D.M.; Murias, M. Estimating the spatial Nyquist of the human EEG. *Behav. Res. Methods* **1998**, *30*, 8–19. [CrossRef]

33. Franken, H.A.; van Strien, J.W.; Franzek, E.J.; van de Wetering, B.J. Error-processing deficits in patients with cocaine dependence. *Biol. Psychol.* **2007**, *75*, 45–51. [CrossRef] [PubMed]

34. Opris, I.; Lebedev, M.A.; Nelson, R.J. Neostriatal neuronal activity correlates better with movement kinematics under certain reward. *Front. Neurosci.* **2016**, *10*, 336. [CrossRef] [PubMed]

35. Levine, T.R.; Hullett, C.R. Eta squared, partial eta squared, and misreporting of effect size in communication research. *Hum. Commun. Res.* **2002**, *28*, 612–625. [CrossRef]

36. Cui, T.; Wang, P.P.; Liu, S.; Zhang, X. P300 amplitude and latency in autism spectrum disorder: A meta-analysis. *Eur. Child Adolesc. Psychiatry* **2017**, *26*, 177–190. [CrossRef] [PubMed]

37. Kemner, C.; Verbaten, M.N.; Cuperus, J.M.; Camfferman, G.; Van Engeland, H. Visual and somatosensory event-related brain potentials in autistic children and three different control groups. *Electroencephalogr. Clin. Neurophysiol.* **1994**, *92*, 225–237. [CrossRef]

38. Kemner, C.; Verbaten, M.N.; Cuperus, J.M.; Camfferman, G.; Van Engeland, H. Auditory event- related potentials in autistic children and three different control groups. *Biol. Psychiatry* **1995**, *38*, 150–165. [CrossRef]

39. Kemner, C.; van der Gaag, R.J.; Verbaten, M.; van Engeland, H. ERP differences among subtypes of pervasive developmental disorders. *Biol. Psychiatry* **1999**, *46*, 781–789. [CrossRef]

40. Donchin, E.; Coles, M.G. Is the p300 component a manifestation of context updating? *Behav. Brain Sci.* **1988**, *11*, 357–374. [CrossRef]

41. Townsend, J.; Westerfield, M.; Leaver, E.; Makeig, S.; Jung, T.; Pierce, K.; Courchesne, E. Event-related brain response abnormalities in autism: Evidence for impaired cerebello-frontal spatial attention networks. *Brain Res. Cogn. Brain Res.* **2001**, *11*, 127–145. [CrossRef]

42. Ciesielski, K.T.; Courchesne, E.; Elmasian, R. Effects of focused attention tasks on event-related potentials in autistic and normal individuals. *Electroencephalogr. Clin. Neurophysiol.* **1990**, *75*, 207–220. [CrossRef]

43. Ferri, R.; Elia, M.; Agarwal, N.; Lanuzza, B.; Musumeci, S.A.; Pennisi, G. The mismatch negativity and the P3a components of the auditory event-related potentials in autistic low-functioning subjects. *Clin. Neurophysiol.* **2003**, *114*, 1671–1680. [CrossRef]

44. Bomba, M.D.; Pang, E.W. Cortical auditory evoked potentials in autism: A review. *Int. J. Psychophysiol.* **2004**, *53*, 161–169. [CrossRef] [PubMed]

45. Verbaten, M.N.; Roelofs, J.W.; van Engeland, H.; Kenemans, J.K.; Slangen, J.L. Abnormal visual event-related potentials of autistic children. *J. Autism Dev. Disord.* **1991**, *21*, 449–470. [CrossRef] [PubMed]

46. Courchesne, E.; Lincoln, A.J.; Yeung-Courchesne, R.; Elmasian, R.; Grillon, C. Pathophysiologic findings in nonretarded autism and receptive developmental disorder. *J. Autism Dev. Disord.* **1989**, *19*, 1–17. [CrossRef] [PubMed]

47. Lincoln, A.J.; Courchesne, E.; Harms, L.; Allen, M. Contextual probability evaluation in autistic, receptive developmental disorder and control children: Event-related potential evidence. *J. Autism Dev. Disord.* **1993**, *23*, 37–58. [CrossRef] [PubMed]

48. Belmonte, M.K.; Yurgelun-Todd, D.A. Functional anatomy of impaired selective attention and compensatory processing in autism. *Brain Res. Cogn. Brain Res.* **2003**, *17*, 651–664. [CrossRef]

49. Belmonte, M.K.; Yurgelun-Todd, D.A. Anatomic dissociation of selective and suppressive processes in visual attention. *Neuroimage* **2003**, *19*, 180–189. [CrossRef]

50. Hopf, J.M.; Vogel, E.; Woodman, G.; Heinze, H.J.; Luck, S.J. Localizing visual discrimination processes in time and space. *J. Neurophysiol.* **2002**, *88*, 2088–2095. [CrossRef] [PubMed]

51. Hillyard, S.A.; Anllo-Vento, L. Event-related brain potentials in the study of visual selective attention. *Proc. Natl. Acad. Sci. USA* **1998**, *95*, 781–787. [CrossRef] [PubMed]

52. Luck, S.J.; Heinze, H.; Mangun, G.R.; Hillyard, S.A. Visual event-related potentials index focused attention within bilateral stimulus arrays. II. Functional dissociation of P1 and N1 components. *Electroencephalogr. Clin. Neurophysiol.* **1990**, *75*, 528–542. [CrossRef]

53. Boutros, N.N.; Korzyukov, O.; Jansen, B.; Feingold, A.; Bell, M. Sensory gating deficits during the mid-latency phase of information processing in medicated schizophrenia patients. *Psychiatry Res.* **2004**, *126*, 203–215. [CrossRef] [PubMed]

54. Casanova, M.F.; Buxhoeveden, D.; Gomez, J. Disruption in the inhibitory architecture of the cell minicolumn: Implications for autism. *Neuroscientist* 2003, *9*, 496–507. [CrossRef] [PubMed]
55. Rubenstein, J.L.; Merzenich, M.M. Model of autism: Increased ratio of excitation/inhibition in key neural systems. *Genes Brain Behav.* 2003, *2*, 255–267. [CrossRef] [PubMed]
56. Overbeek, T.J.M.; Nieuwenhuis, S.; Ridderinkhof, K.R. Dissociable components of error processing. *J. Psychophysiol.* 2005, *19*, 319–329. [CrossRef]
57. Yeung, N.; Botvinick, M.M.; Cohen, J.D. The neural basis of error detection: Conflict monitoring and the error-related negativity. *Psychol. Rev.* 2004, *111*, 931–959. [CrossRef] [PubMed]
58. Yeung, N.; Cohen, J.D. The impact of cognitive deficits on conflict monitoring. Predictable dissociations between the error-related negativity and N2. *Psychol. Sci.* 2006, *17*, 164–171. [CrossRef] [PubMed]
59. Van Veen, V.; Carter, C.S. The timing of action-monitoring process in the anterior cingulate cortex. *J. Cogn. Neurosci.* 2002, *14*, 593–602. [CrossRef] [PubMed]
60. Burle, B.; Vidal, F.; Tandonnet, C.; Hasbroucq, T. Physiological evidence for response inhibition in choice reaction time tasks. *Brain Cogn.* 2004, *56*, 153–164. [CrossRef] [PubMed]
61. Ridderinkhof, K.R.; Ullsperger, M.; Crone, E.A.; Nieuwenhuis, S. The role of the medial frontal cortex in cognitive control. *Science* 2004, *306*, 443–447. [CrossRef] [PubMed]
62. Casanova, M.F. Minicolumnar pathology in autism. In *Recent Developments in Autism Research*; Casanova, M.F., Ed.; Nova Biomedical Books: New York, NY, USA, 2005; pp. 133–143, ISBN 1-59454-497-2.
63. Casanova, M.F. Neuropathological and genetic findings in autism: The significance of a putative minicolumnopathy. *Neuroscientist* 2006, *12*, 435–441. [CrossRef] [PubMed]
64. Just, M.A.; Cherkassky, V.; Keller, T.A.; Minshew, N.J. Cortical activation and synchronization during sequence comprehension in high-functioning autism: Evidence of underconnectivity. *Brain* 2004, *127*, 1811–1821. [CrossRef] [PubMed]
65. Williams, E.L.; Casanova, M.F. Autism and dyslexia: A spectrum of cognitive styles as defined by minicolumnar morphometry. *Med. Hypotheses* 2009, *74*, 59–62. [CrossRef] [PubMed]
66. Bogte, H.; Flamma, B.; van der Meere, J.; van Engeland, H. Post-error adaptation in adults with high functioning autism. *Neuropsychologia* 2007, *45*, 1707–1714. [CrossRef] [PubMed]
67. Bauman, M.L.; Kemper, T.L. Structural brain anatomy in autism: What is the evidence? In *The Neurobiology of Autism*, 2nd ed.; Bauman, M.L., Kemper, T.L., Eds.; John Hopkins University Press: Baltimore, MD, USA, 2005; pp. 121–135, ISBN 0801880475.
68. Minshew, N.J.; Sweeney, J.A.; Bauman, M.L.; Webb, S.J. Neurological aspects of autism. In *Handbook of Autism and Pervasive Developmental Disorders, Diagnosis, Development, Neurobiology, and Behavior*, 3rd ed.; Volkmar, F.R., Paul, R., Klin, A., Cohen, D., Eds.; Wiley: New York, NY, USA, 2005; Volume 1, pp. 473–514, ISBN 0471716960.
69. Mundy, P. Annotation: The neural basis of social impairments in autism: The role of the dorsal medial-frontal cortex and anterior cingulate system. *J. Child Psychol. Psychiatry* 2003, *44*, 793–809. [CrossRef] [PubMed]
70. Henderson, H.; Schwartz, C.; Mundy, P.; Burnette, C.; Sutton, S.; Zahka, N.; Pradella, A. Response monitoring, the error-related negativity, and differences in social behavior in autism. *Brain Cogn.* 2006, *61*, 96–109. [CrossRef] [PubMed]
71. Devinsky, O.; Luciano, D. The contributions of cingulate cortex to human behavior. I. In *Neurobiology of Cingulate Cortex and Limbic Thalamus: A Comprehensive Handbook*; Gabriel, M., Vogt, B.A., Eds.; Birkhauser: Boston, MA, USA, 1993; pp. 527–556, ISBN 978-1-4899-6706-0.
72. Tamnes, C.K.; Walhovd, K.B.; Torstveit, M.; Sells, V.T.; Fjell, A.M. Performance monitoring in children and adolescents: A review of developmental changes in the error-related negativity and brain maturation. *Dev. Cogn. Neurosci.* 2013, *6*, 1–13. [CrossRef] [PubMed]

behavioral sciences

MDPI

Article

Rotating and Neurochemical Activity of Rats Lesioned with Quinolinic Acid and Transplanted with Bone Marrow Mononuclear Cells

Teresa Serrano Sánchez [1,*], María Elena González Fraguela [1], Lisette Blanco Lezcano [2], Esteban Alberti Amador [3], Beatriz Caballero Fernández [4], María de los Ángeles Robinson Agramonte [1], Lourdes Lorigados Pedre [1] and Jorge A Bergado Rosado [2]

[1] Immunochemical Department, International Center for Neurological Restoration, 25th Ave, Playa, 15805, Havana PC 11300, Cuba; marie@neuro.ciren.cu (M.E.G.F.); robin@neuro.ciren.cu (M.d.l.Á.R.A.); lourdesl@neuro.ciren.cu (L.L.P.)
[2] Experimental Neurophysiology Department, International Center of Neurological Restoration (CIREN) Ave. 25 No. 15805 e/158 and 160, Playa, Havana 11300, Cuba; lblanco@neuro.ciren.cu (L.B.L.); jorgebergado@yahoo.com (J.A.B.R.)
[3] Molecular biology Department, International Center of Neurological Restoration (CIREN) Ave. 25 No. 15805 e/158 and 160, Playa, Havana 11300, Cuba; alberti@neuro.ciren.cu
[4] Policlínico 26 de Julio, Calle 72 #1313. A. Almendares, Playa, Havana 11300, Cuba; bettycfdez@infomed.sld.cu
* Correspondence author: teresa@neuro.ciren.cu; Tel.: +53-7-2715353

Received: 10 August 2018; Accepted: 17 September 2018; Published: 20 September 2018

Abstract: Huntington's disease (HD) is an inherited, neurodegenerative disorder that results from the degeneration of striatal neurons, mainly GABAergic neurons. The study of neurochemical activity has provided reliable markers to explain motor disorders. To treat neurodegenerative diseases, stem cell transplants with bone marrow (BM) have been performed for several decades. In this work we determine the effect of mononuclear bone marrow cell (mBMC) transplantation on the rotational behavior and neurochemical activity in a model of Huntington's disease in rats. Four experimental groups were organized: Group I: Control animals ($n = 5$); Group II: Lesion with quinolinic acid (QA) in the striatum ($n = 5$); Group III: Lesion with QA and transplant with mBMC ($n = 5$); Group IV: Lesion with QA and transplant with culture medium (Dulbecco's modified Eagle's medium (DMEM) injection) ($n = 5$). The rotational activity induced by D-amphetamine was evaluated and the concentration of the neurotransmitter amino acids (glutamate and GABA) was studied. The striatal cell transplantation decreases the rotations induced by D-amphetamine ($p < 0.04$, Wilcoxon matched pairs test) and improves the changes produced in the levels of neurotransmitters studied. This work suggests that the loss of GABAergic neurons in the brain of rats lesioned with AQ produces behavioral and neurochemical alterations that can be reversed with the use of bone marrow mononuclear cell transplants.

Keywords: striatum; quinolinic acid; transplantation; mononuclear cell; bone marrow cell; Huntington disease model

1. Introduction

Huntington's disease (HD) has been associated with a degeneration of striatal cells, mainly death of medium spiny GABAergic projection neurons within the caudate nucleus and putamen. Although the most pronounced pathology is observed in the basal ganglia, cell death also occurs early in the cerebral cortex [1]. The experimental models play essential roles in the understanding of disease mechanisms, progression, and drug efficacy testing. The injection of quinolinic acid into the striatum is used as a theoretical model of HD [2]. Quinolinic acid (QA) produces loss of GABAergic neurons; in this way there is an imbalance between the healthy and damaged striatum in response to the action of an intact dopaminergic system, which produces an asymmetric motor response to the action of a dopaminergic agonist, D-amphetamine.

The study of the neurochemical activity in the affected brain tissue is a reliable marker to explain the motor disorders that appear in this disease [3]. No therapy has been shown to delay disease onset or slow progression in humans. Unlike other neurodegenerative entities, pharmacological treatment is only a palliative procedure. Thus, there is an urgent need to identify and validate other treatments. Cells isolated from bone marrow are successfully used for transplantation in experimental models of cranial encephalic trauma [4,5] as well as in striatal ischemia to reduce the motor deficit that appears after damage [6]. Recently, autologous bone marrow stem cell transplantation has been used to reverse the cognitive deficit observed in an experimental model of Huntington's disease, which demonstrates that bone marrow cell transplantation is able to reduce the cognitive damage that appears in this model [7]. In general, the repair mechanism consists of the potential to develop into many different types of cells (thus serving as a repair system for the body) and the potential to release trophic factors that repair and regulate other compensatory mechanisms in the damaged area. For this reason, in the last decades, neurorestorative treatment has been tried in experimental models that use stem cells from different sources, such as mononuclear bone marrow cells (mBMC) [8,9].

In the present study, the objective was to determine the effect of mBMC transplantation on the rotational activity induced by systemic administration of D-amphetamine (5 mg/kg intraperitoneal) and the neurochemical activity in two brain areas: the cerebral cortex and striatum. We show evidence that indicates the transplant may be a viable therapeutic option for Huntington's disease.

2. Materials and Methods

2.1. Animals

Adult male Sprague Dawley (SD) rats (total $N = 20$) obtained from the National Center for Laboratory Animals Production (CENPALAB) and weighing 200–250 g (between 9 and 12 weeks of age) were used in this study. Animals were housed in translucent makrolon cages (five animals per cage) under a 12 h light: 12 h darkness cycle with ad libitum access to water and food.

For the behavior and neurotransmission study, four experimental groups were included ($n = 5$ each): Control (C), Group I; QA lesion (QAL), Group II; QA lesion + mononuclear bone marrow cell transplantation (QA+mBMC), Group III; and QA lesion + Dulbecco's modified Eagle's medium (QA+DMEM) injection, Group VI.

2.2. Striatal QA Lesions

Unilateral lesions of the right striatum were produced by the intrastriatal injection of QA. Rats were anesthetized with an intraperitoneal (ip) injection of ketamine (50 mg/mL, ip, IMEFA, Havana, Cuba). QA injections were administered with the help of a stereotaxic apparatus (model 900; David Kopf Instruments, Tujunga, CA, USA). Each rat was injected in the striatum with 1.2 μL of QA (125.5 nmol) (Sigma, Saint Louis, MA, USA) using a 30 G Hamilton syringe at the following coordinates: 1.2 mm anterior and 2.8 mm lateral to the bregma, and 5.5 mm below the dura. The toxin was injected over a period of 1 min, and the cannula was left in place for an additional 10 min before being slowly removed.

2.3. Obtaining Rat Mononuclear Bone Marrow Cells

The mBMC were isolated from the rat femur as described in the work of Woodbury and colleagues [10]. Male SD rats aged between 32 and 48 days old were anesthetized with an intraperitoneal injection of ketamine (50 mg/mL, ip, IMEFA, Havana, Cuba) and a cut on the skin of the hind limbs was performed, separating the tissue parallel to the bone to extract both femurs; then the animals were euthanized with a lethal overdose of chloral hydrate. The extracted bones were placed for 30 min on a Petri dish containing 0.9% physiologic saline, after which the bone marrow was obtained by flushing with sterile phosphate-buffered saline (PBS; NaCl, 8 g/L; KCl, 0.2 g/L; Na$_2$HPO$_4$, 1.09 g/L; KH$_2$PO$_4$, 0.26 g/L, pH 7.2) through one of the femoral epiphyses. The bone marrow was collected in sterile containers to be later washed by centrifugation.

2.4. Isolation of Mononuclear Bone Marrow Cells

The suspension of bone marrow cells was washed three times with 1× PBS by centrifugation for 10 min at 2000 rpm at 20 °C. An aliquot of 2.5 mL of Ficoll–Hypaque was placed on the bottom of a graduated glass tube, on top of which 5 mL of the cellular suspension in PBS was layered. This was centrifuged for 45 min at 2800 rpm at 20 °C.

The mononuclear cell band was extracted with a pipette and washed immediately, discarding the supernatant into a container with hypochlorite and collecting the cellular pellet, which was suspended in 1× PBS and cell viability was determined by trypam blue exclusion.

2.5. Transplantation

Four weeks after the QA lesion, the animals to be transplanted with mBMCs were deeply anesthetized as previously described. Rats were placed in the stereotactic apparatus and the skin over the skull was reopened. Using a Hamilton syringe, the mBMC suspension (50,000 cells/μL in DMEM) was injected into the lesioned striatum (in two deposits; 1 μL per deposit) at coordinates slightly different from the ones used for the QA lesions: 0.7 mm anterior from the bregma, 2.8 mm lateral from the midline, and 5.5 and 4.6 mm under the dura surface. Cells were injected slowly over a period of 1 min. The needle was left in place for an additional 10 min following injection and then carefully removed. Sham-transplanted animals (DMEM) received an equal volume of tissue culture medium injected in the same way, at the same stereotactic coordinates.

2.6. Behavioral Tests (Rotating Activity Induced by D-Amphetamine)

The rotational activity induced by D-amphetamine (5 mg/kg, ip, Sigma, St. Luis, MO, USA) [11] was studied one week after the QA injury and one month after the mBMC/DMEM transplant, using an LE 3806 Electronic Multicounter coupled to LE 902 sensors (PanLAB, Barcelona, Spain) that measured the sense of rotation. The measurements were performed for a period of 90 min.

2.7. Obtaining Samples for Neurotransmission Study

After the behavioral test was over, the rats received an overdose of ketamine (100 mg/mL, ip, IMEFA, Havana, Cuba) and were decapitated. Their brains were extracted and washed with cold 0.9% NaCl, after which the striatum (St) and prefrontal cortex was dissected. The obtained tissue was frozen in liquid nitrogen, weighed and stored at −80 °C for further analysis.

2.8. Neurotransmitter Study

The amino acid concentrations in the tissue were determined by high-performance liquid chromatography (HPLC), coupled to a fluorescence detector and using derivatization with o-phthaldialdehyde (OPA). Then, 10 μL of sample was mixed with 10 μL of the OPA derivatizing agent (10 mM OPA dissolved in 0.1 mol/L sodium tetraborate buffer containing 77 mM of 3-mercaptopropionic acid and 10% methanol pH 9.3). The mixture was vortexed for 15 s and the reaction was stopped with 5% acetic acid at 45 s. From this mixture, 20 μL was injected to the chromatograph with a Hamilton syringe. The derivatized amino acids were passed through a reverse phase column (HR-80, 8 cm long with a4.6-mm internal diameter, ESA), with a similar stationary phase precolumn, by means of an isocratic chromatographic pump (Philips PU, Amsterdam, The Netherlands) 4100) and were detected by a fluorescence detector with excitation λ = 340 nm and emission λ = 460 nm (Philips PU 4027).

The chromatograms were recorded using the program CHROMATEPC version 4.24 (Philips). A mobile phase composed of 0.1 mol/L NaH_2PO_4 and 20% methanol was used to separate the amino acids.

2.9. Ethical Considerations

Experiments were carried out in accordance with the Cuban Regulations for the Use of Laboratory Animals (CENPALAB 1997) and the Canadian Council on Animal Care (CCAC) [12] This work was approved by the Ethical Committee of the International Center for Neurological Restoration. Efforts were made to minimize the pain and discomfort of the animals, as well as the number of animal used for experiments.

2.10. Statistical Processing

Statistical analysis was carried out using Statistical software. The values are expressed as mean $\pm$ SEM. The normal distribution and homogeneity of variance of the data were tested by the Kolmogorov–Smirnov and Levene tests, respectively. The comparisons between more than two groups were made by one-way non parametric ANOVA; the Kruskall–Wallis test was applied to aminoacid analysis, and the Wilcoxon test was applied to behavioral evaluations. In all cases, statistically significant differences were considered when $p \leq 0.05$.

3. Results

Our results confirm the hypothesis that the transplantation of bone marrow mononuclear cells in the lesioned striatum of rats reverts the neurochemical alterations that appear in the neurotoxic model of Huntington's disease induced by quinolinic acid. The objective was to evaluate the possible protective effect of transplanted mBMC on the changes induced in motor behavior and neurotransmission in the HD model.

3.1. Rotational Activity

Figure 1 show the ipsilateral to the lesion rotatory activity exhibited by the animals injected with QA in striatum under D-amphetamine effects. The striatal transplant of mBMC diminished the rotation induced by D-amphetamine.

Rotary activity induced by D-Amphetamine

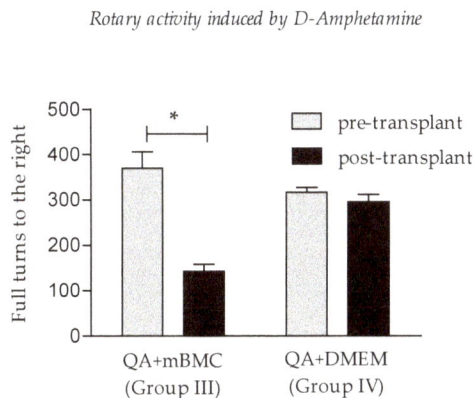

Figure 1. Rotatory activity induced under D-amphetamine. Lesioned group with quinolinic acid + mononuclear bone marrow cell transplant (QA+mBMC, Group III) and lesioned group with QA+Dulbecco's modified Eagle's medium (DMEM) transplant (QA+DMEM, Group IV). Abscissa, number of complete right turns ipsilateral to the lesioned hemisphere. Data are expressed as mean $\pm$ standard error of mean (SEM). Ordinates: experimental groups (* equivalent $p \leq 0.04$).

3.2. Neurochemical Activity

3.2.1. Glutamate Concentration

In the Figure 2 the comparison of the glutamate content in the experimental groups for the right striatum (RS) and right cortex (RC) showed significant differences among them (RS $p \leq 0.001$ and RC $p \leq 0.001$). The higher glutamate content was obtained for Group III (QA+mBMC) for both structures.

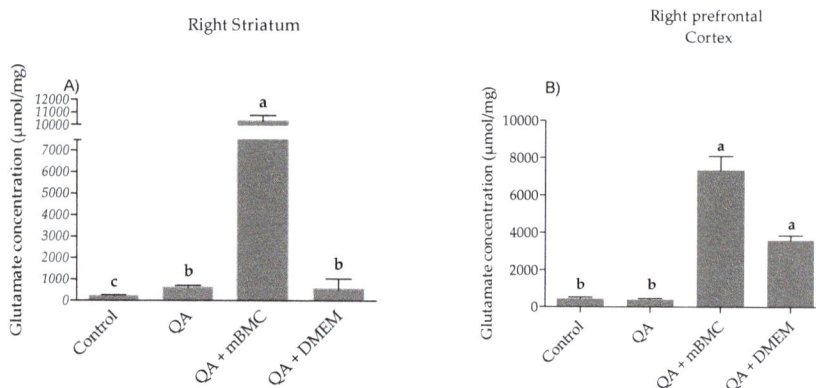

Figure 2. Study of glutamate (GLU) concentration in lesioned rats with transplanted mBMC. (**A**) Comparison of GLU concentration in the right striatum among the experimental groups: H (3, 20) = 16.28, $p < 0.001$. (**B**) GLU concentration in right prefrontal cortex among the experimental groups: H (3, 19) = 15.53, $p < 0.001$. Data are expressed as mean $\pm$ SEM. The letters in the top the bar correspond to statistical differences among the experimental groups. The data analysis was carried out through one-way non parametric ANOVA, using the Kruskall–Wallis test (common letters: non-significant differences; different letters; significant differences).QA lesion (Group II); lesioned group with QA + transplant (QA+mBMC, Group III); and lesioned group with QA+DMEM transplant (QA+DMEM, Group IV).

3.2.2. GABA (γ-Aminobutyric Acid) Concentration

Finally, in Figure 3 the comparison of GABA (content in the right striatum (RS) and right cortex (RC) among the experimental groups showed significant differences (RS and RC: ($p \leq 0.001$, the higher GABA content was obtained in both structures for Group III in comparison to the control group.

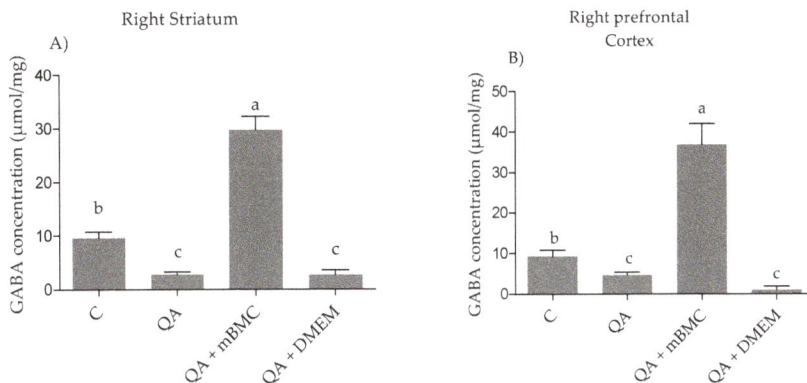

Figure 3. Study of GABA (γ-aminobutyric acid) concentration in lesioned rats + mBMC transplanted. (**A**) Comparison of GABA concentration in the right striatum among experimental groups: H (3, 18) = 12.55, ($p \leq 0.005$). (**B**) GABA concentration in right prefrontal cortex among experimental group: H (3, 15) = 12.72, ($p \leq 0.005$). Data are expressed as mean ± SEM. The letters at the top the bars correspond to statistical differences among experimental groups. The data analysis was carried out through one-way non parametric ANOVA, using the Kruskall–Wallis test (common letters: non-significant differences; different letters; significant differences). QA lesion (Group II); (lesioned group with QA + transplant (QA+mBMC, Group III); and lesioned group with QA+DMEM transplant (QA+DMEM, Group IV).

4. Discussion

It has been suggested that striatum is the most affected structure in Huntington's disease, with major relevance of the spiny neurons of medium size which integrate the nucleus and main receptors of the most important connections with the cerebral cortex [13,14]. The striatum is formed by different types of neural cells, but medium-sized spiny neurons, which use GABA as a neurotransmitter, represent 90–95% of the neurons found and are its main projection cells [15]. The rest of the neuronal types of the striatum are less frequent and include neurons with long dendrites that function as interneurons and use acetylcholine as neurotransmitter. The degeneration of GABAergic cells of medium size of the striatum causes the loss of influence of the neurotransmitter GABA on their targets, affecting by different pathways the "direct" and "indirect" transmission network of the motor circuit [16–18].

In this study, the rotational activity induced by D-amphetamine was evaluated as predictive of the degree of motor deficit present in the rats. This analysis was performed either before or after the mBMC transplant. This dose-dependent motor behavior is usually accompanied by episodes of "rotations" [19,20], which are attributed to an asymmetry of dopamine activity between the lesioned and healthy striatum [21]. The typical stereotyped behavior shown by the healthy group represents a hyperkinetic reaction characteristic of an intact brain before the action of a dopaminergic agonist (data not shown). Hence, the results of this test suggest that the damaged rats developed a rotating activity ipsilateral to the lesioned hemisphere, probably associated with the loss of GABAergic projections, as has already been described by other authors [19].

On the other hand, the rotational activity induced by D-amphetamine decreased significantly in the transplant group, maintaining a certain degree of rotational activity justified by the stereotypic activity caused by D-amphetamine administration. The dopaminergic imbalance observed in the present study does not respond to the decline in the dopamine synthesis process; this dopaminergic imbalance could be related to the neuronal loss of GABAergic striatal neurons which receive the nigrostriatal dopaminergic projections [21]. It is important to emphasize the fact that the rotational behavior induced by D-amphetamine in our study is only an approximate and indirect way of studying neurodegeneration in Huntington's disease. This test only indicates the motor asymmetry that occurs in response to an imbalance in the striatum function product of the ipsilateral lesion by QA.

In this sense, previous data from our group showed that the transplant has a trophic action on the damaged tissue [22], which suggests a mechanism by which the mBMC transplant reduces the dopaminergic asymmetry between the hemispheres. The trophic support could justify the effectiveness of therapy in this area on the recovery of damaged cells, which is in line with authors agreeing on the effectiveness of mBMC transplant for improving behavior in rats lesioned with quinolinic acid [19,23].

The mechanisms by which the transplanted cells produce the recovery of normal neurochemistry in the brain are multifactorial. There is a previous report [22] with the observation that transplant cells express the NeuN protein, suggesting the transformation of them into a GABAergic neural phenotype. In another sense it is possible that the arrival of BDNF (Brain Derived Neurotrophic Factor) from other areas to the lesioned tissue contributes to the positive effect observed after the transplant. It is known that the lesion striatum can stimulate neurogenesis in the subventricular zone and produce trophic factors, including BDNF [24]. However, the possibility that the release of BDNF detected in the study reported by us [22] and that, among other actions, produces an increase in GAD (Glutamate decarboxylase), comes from the cells that are grafted and linked to the microenvironment is not excluded. On the other hand, the transplant offers an appropriate scenario to produce a positive effect in the lesioned area through the activation of other factors such as interleukins and/or growth factors. These agents guarantee the survival of nerve cells, and raise the threshold of resistance to quinolinic acid. Also, these cells are part of the immune system; protective autoimmunity effects are described as a physiological response that occurs after damage [25] that depends directly on the microenvironment where they were implanted.

Although reports of the activity of GABA and glutamate are described in experimental models [26], there is no information available on the changes in the concentration of these amino acids in the experimental model of Huntington's disease induced by quinolinic acid and the effectiveness of mBMC transplantation.

Glutamatergic synapses are the most abundant neurotransmission systems in the central nervous system of mammals [27]. The functional complexity of these synapses is reflected at the presynaptic level by the existence of multiple excitotoxicity mechanisms initiated by the activation of autoreceptors which respond well to glutamate itself or to heterorreceptors that enhance or inhibit the release of the neurotransmitter.

The striatum begins to degenerate before other brain areas, and altered activity at corticostriatal synapses contributes to an imbalance in survival versus death signaling pathways in this brain region. Striatum projection neurons of the indirect pathway are the most vulnerable, and their dysfunction contributes to motor symptoms occurring at early stages of the disease [28]. There is evidence of expression changes in striatal excitatory synaptic activity by decreasing glutamate uptake and increasing signaling at N-methyl-d-aspartate receptors (NMDAR) [29].

In this study, we show the tissue concentration of the neurotransmitters as indicative of synthesis and storage processes. In addition, in the case of the striatum, this study reflects a release process because this structure is the target of cortico-striatal glutamatergic innervation. With this argument, we support the hypothesis that the increase in the glutamate content observed in the transplanted striatum is the result of all the tissue content of the neurotransmitters. Hence, the increase in glutamate in striatum and cortex observed in our results may not be deleterious for these neuronal populations, since the result of the behavior after the transplant was beneficial. More studies must be carried out to clarify this hypothesis. On the other hand, it is observed that the DMEM alone causes an induction of glutamate release. This was an expected result considering that as DMEM contains high levels of amino acids, vitamins, and glucose, it would not be surprising for some benefit to be reported after the local application of it in the lesioned tissue. For example, and in accordance with this result, beneficial effects are reported in the literature with the injection of DMEM on the maintenance and differentiation of epithelial cells [30]. However, none of these effects is as potent as that demonstrated by the mBMC transplant.

Behav. Sci. **2018**, *8*, 87

The decrease in the concentration of GABA and glutamate in lesioned animals might be related with the injury quinolinic acid and/other eventual loss of GABAergic cells [31–33]. Since several studies show that the administration of the neurotoxin induces cytotoxic effects on various types of neurons, which could be possibly occur though glutamate and GABA recapture systems not being intact or by damage to the blood–brain barrier. In addition, the cell transplantation could produce dendritic growth associated with the appearance of intrinsic GABAergic neuronal circuits. If we take into account the abovementioned information, it is possible that transplantation is an important factor in the establishment of the architecture and the interneuronal communication. Consequently, it modifies the physiology of the synapses of GABAergic type influencing the neurotrasmition system.

GABAergic interneurons play a prominent role in the function of the cerebral cortex since they allow the synchronization of pyramidal neurons and greatly influence their differentiation and maturation during development [34]. A certain retrograde transport of quinolinic acid striatum–cortex with deleterious effects on the interneurons of that region may justify the decrease in GABA found in our study, although future studies must be conducted to corroborate this argument.

In this regard, changes in amino acids are likely to be the product initially of loss of GABAergic cells induced by the lesion at first stage. Previous studies conducted by our group show that transplanted cells survive for a prolonged period of time, producing a trophic action at the implant site [35]. It is also likely that this occurs after the transplant as result of the trophic action of the transplanted cells. In addition, in previous studies, we demonstrated the histopathological changes that occur in the brain lesioned with quinolinic acid and transplanted with mBMCs, through morphological and immunohistochemical studies [36,37]. The observation using cresyl violet allowed a general inspection of the striatal morphology in the different experimental groups, showing the normal cell distribution in this region. QA causes neuronal death and structural alterations, drastically reducing cell density and disrupting the pattern of cell distribution. On the other hand, our group demonstrated through an immunofluorescence study with GFAP and FJC (Fluoro-Jade C) staining that there was intense reactivity for GFAP [31,35] in the striatum of lesioned and DMEM-treated animals, indicating astrocytic gliosis in the tissue. Such a reaction was absent in the striatum of control and transplanted animals. A similar result was seen after marking with FJC, which stains degenerating neurons. Lesioned and DMEM-treated animals showed evidence of intense degeneration, which was not seen in control and transplanted animals. Finally, they were studied through immunofluorescence study by NeuN and FJC staining [36,37] whereby NeuN, a neuronal marker, showed positive reactivity in the control and the transplanted groups. In contrast, these groups were negative for FJC. The merged panel shows positive cellular bodies for NeuN in control and mBMC-transplanted groups, suggesting an integration of transplanted cells and even raising the possibility of neuronal differentiation among the transplanted cells, an aspect that we should consider in future experiments.

5. Conclusions

In the present study, we demonstrated that the loss of GABAergic neurons in the rat striatum as a consequence of AQ injection produces neurochemical and behavioral alterations that can be reversed with a neurorestorative treatment based on the transplantation of mononuclear bone marrow cells. Further studies will be required to understand the mechanisms by which the transplanted cells correct behavioral and neurochemical alterations produced in this model, a novel contribution of this paper.

Author Contributions: T.S.S. Conceived and designed the experiments, performed the review of the literature, and wrote the manuscript. M.E.G.F. participated in the review of the literature and in the study of amino acid determination. L.B.L. participated in the analysis of the data and the writing of the manuscript. E.A.A., B.C.F., M.d.l.Á.R.A., L.L.P and J.A.B.R. participated in the analysis of the data and revision of the manuscript.

Funding: This research was funded by International Center for Neurological Restoration (CIREN). Ave 25 No. 15805 entre 158 y 159, Cubanacan, Havana, Cuba.

Acknowledgments: This study was supported financially by the CIREN Scientific Council.

Conflicts of Interest: The authors declare no conflict of interest.

References

1. Dayalu, P.; Albin, R.L. Huntington disease: Pathogenesis and treatment. *Neurol. Clin.* **2015**, *33*, 101–114. [CrossRef] [PubMed]
2. Davies, S.; Ramsden, D.B. Huntington's disease. *Mol. Pathol.* **2001**, *54*, 409–413. [CrossRef] [PubMed]
3. Araujo, D.M.; Hilt, D.C. Glial cell line-derived neurotrophic factor attenuates the excitotoxin-induced behavioral and neurochemical deficits in a rodent model of Huntington's disease. *Neuroscience* **1997**, *81*, 1099–1110. [CrossRef]
4. Cox, C.S., Jr.; Hetz, R.A.; Liao, G.P.; Aertker, B.M.; Ewing-Cobbs, L.; Juranek, J.; Savitz, S.I.; Jackson, M.L.; Romanowska-Pawliczek, A.M.; Triolo, F.; et al. Treatment of Severe Adult Traumatic Brain Injury Using Bone Marrow Mononuclear Cells. *Stem Cells* **2017**, *35*, 1065–1079. [CrossRef] [PubMed]
5. Hasan, A.; Deeb, G.; Rahal, R.; Atwi, K.; Mondello, S.; Marei, H.E.; Gali, A.; Sleiman, E. Mesenchymal Stem Cells in the Treatment of Traumatic Brain Injury. *Front Neurol.* **2017**, *8*, 28. [CrossRef] [PubMed]
6. Chen, J.; Yang, Y.; Shen, L.; Ding, W.; Chen, X.; Wu, E.; Cai, K.; Wang, G. Hypoxic Preconditioning Augments the Therapeutic Efficacy of Bone Marrow Stromal Cells in a Rat Ischemic Stroke Model. *Cell. Mol. Neurobiol.* **2017**, *37*, 1115–1129. [CrossRef] [PubMed]
7. Lescaudron, L.; Unni, D.; Dunbar, G.L. Autologous adult bone marrow stem cell transplantation in an animal model of huntington's disease: Behavioral and morphological outcomes. *Int. J. Neurosci.* **2003**, *113*, 945–956. [CrossRef] [PubMed]
8. Kerkis, I.; Haddad, M.S.; Valverde, C.W.; Glosman, S. Neural and mesenchymal stem cells in animal models of Huntington's disease: Past experiences and future challenges. *Stem Cell Res. Ther.* **2015**, *6*, 232. [CrossRef] [PubMed]
9. Rossignol, J.; Fink, K.D.; Crane, A.T.; Davis, K.K.; Bombard, M.C.; Clerc, S.; Bavar, A.M.; Lowrance, S.A.; Song, C.; Witte, S.; et al. Reductions in behavioral deficits and neuropathology in the R6/2 mouse model of Huntington's disease following transplantation of bone-marrow-derived mesenchymal stem cells is dependent on passage number. *Stem Cell Res. Ther.* **2015**, *6*, 9. [CrossRef] [PubMed]
10. Woodbury, D.; Schwarz, E.J.; Prockop, D.J.; Black, I.B. Adult rat and human bone marrow stromal cells differentiate into neurons. *J. Neurosci. Res.* **2000**, *61*, 364–370. [CrossRef]
11. Ungerstedt, U.; Arbuthnott, G.W. Quantitative recording of rotational behavior in rats after 6-hydroxy-dopamine lesions of the nigrostriatal dopamine system. *Brain Res.* **1970**, *24*, 485–493. [CrossRef]
12. National Research Council (US). *National Research Council (US) Committee on Guidelines for the Care and Use of Animals in Neuroscience and Behavioral Research*; National Academies Press: Washington, DC, USA, 2003; pp. 1–224.
13. DeLong, M.; Wichmann, T. Changing views of basal ganglia circuits and circuit disorders. *Clin. EEG Neurosci.* **2010**, *41*, 61–67. [CrossRef] [PubMed]
14. DeLong, M.R.; Wichmann, T. Basal Ganglia Circuits as Targets for Neuromodulation in Parkinson Disease. *JAMA Neurol.* **2015**, *72*, 1354–1360. [CrossRef] [PubMed]
15. Ribak, C.E.; Vaughn, J.E.; Roberts, E. The GABA neurons and their axon terminals in rat corpus striatum as demonstrated by GAD immunocytochemistry. *J. Comp. Neurol.* **1979**, *187*, 261–283. [CrossRef] [PubMed]
16. Bergman, H.; Feingold, A.; Nini, A.; Raz, A.; Slovin, H.; Abeles, M.; Vaadia, E. Physiological aspects of information processing in the basal ganglia of normal and parkinsonian primates. *Trends Neurosci.* **1998**, *21*, 32–38. [CrossRef]
17. Kelly, R.M.; Strick, P.L. Macro-architecture of basal ganglia loops with the cerebral cortex: Use of rabies virus to reveal multisynaptic circuits. *Prog. Brain Res.* **2004**, *143*, 449–459. [PubMed]
18. Obeso, J.A.; Rodriguez-Oroz, M.C.; Rodriguez, M.; Lanciego, J.L.; Artieda, J.; Gonzalo, N.; Olanow, C.W. Pathophysiology of the basal ganglia in Parkinson's disease. *Trends Neurosci.* **2000**, *23*, S8–S19. [CrossRef]
19. Hantraye, P.; Riche, D.; Maziere, M.; Isacson, O. A primate model of Huntington's disease: Behavioral and anatomical studies of unilateral excitotoxic lesions of the caudate-putamen in the baboon. *Exp. Neurol.* **1990**, *108*, 91–104. [CrossRef]
20. Schwarcz, R.; Fuxe, K.; Agnati, L.F.; Hokfelt, T.; Coyle, J.T. Rotational behaviour in rats with unilateral striatal kainic acid lesions: A behavioural model for studies on intact dopamine receptors. *Brain Res.* **1979**, *170*, 485–495. [CrossRef]

21. Fuxe, K.; Janson, A.M.; Jansson, A.; Andersson, K.; Eneroth, P.; Agnati, L.F. Chronic nicotine treatment increases dopamine levels and reduces dopamine utilization in substantia nigra and in surviving forebrain dopamine nerve terminal systems after a partial di-mesencephalic hemitransection. *Naunyn Schmiedebergs Arch. Pharmacol.* **1990**, *341*, 171–181. [CrossRef] [PubMed]

22. Serrano, S.T.; Alberti, A.E.; Lorigados, P.L.; Blanco, L.L.; Diaz, A.I.; Bergado, J.A. BDNF in quinolinic acid lesioned rats after bone marrow cells transplant. *Neurosci. Lett.* **2014**, *559*, 147–151. [CrossRef] [PubMed]

23. Rossignol, J.; Fink, K.; Davis, K.; Clerc, S.; Crane, A.; Matchynski, J.; Lowrance, S.; Bombard, M.; Dekorver, N.; Lescaudron, L.; et al. Transplants of adult mesenchymal and neural stem cells provide neuroprotection and behavioral sparing in a transgenic rat model of Huntington's disease. *Stem Cells* **2014**, *32*, 500–509. [CrossRef] [PubMed]

24. Tattersfield, A.S.; Croon, R.J.; Liu, Y.W.; Kells, A.P.; Faull, R.L.; Connor, B. Neurogenesis in the striatum of the quinolinic acid lesion model of Huntington's disease. *Neuroscience* **2004**, *127*, 319–332. [CrossRef] [PubMed]

25. Del Barco, D.G.; Berlanga, J.; Penton, E.; Hardiman, O.; Montero, E. Boosting controlled autoimmunity: A new therapeutic target for CNS disorders. *Expert Rev. Neurother.* **2008**, *8*, 819–825. [CrossRef] [PubMed]

26. Haber, S.N. Neurotransmitters in the human and nonhuman primate basal ganglia. *Hum. Neurobiol.* **1986**, *5*, 159–168. [PubMed]

27. Moberly, A.H.; Czarnecki, L.A.; Pottackal, J.; Rubinstein, T.; Turkel, D.J.; Kass, M.D.; McGann, J.P. Intranasal exposure to manganese disrupts neurotransmitter release from glutamatergic synapses in the central nervous system in vivo. *Neurotoxicology* **2012**, *33*, 996–1004. [CrossRef] [PubMed]

28. Van, D.E.; Reedeker, N.; Giltay, E.J.; Eindhoven, D.; Roos, R.A.; van der Mast, R.C. Course of irritability, depression and apathy in Huntington's disease in relation to motor symptoms during a two-year follow-up period. *Neurodegener. Dis.* **2014**, *13*, 9–16.

29. Lau, A.; Tymianski, M. Glutamate receptors, neurotoxicity and neurodegeneration. *Pflugers Arch.* **2010**, *460*, 525–542. [CrossRef] [PubMed]

30. Ahmado, A.; Carr, A.J.; Vugler, A.A.; Semo, M.; Gias, C.; Lawrence, J.M.; Chen, L.L.; Chen, F.K.; Turowski, P.; da Cruz, L.; Coffey, P.J. Induction of differentiation by pyruvate and DMEM in the human retinal pigment epithelium cell line ARPE-19. *Investig. Ophthalmol. Vis. Sci.* **2011**, *52*, 7148–7159. [CrossRef] [PubMed]

31. Braidy, N.; Grant, R.; Adams, S.; Brew, B.J.; Guillemin, G.J. Mechanism for quinolinic acid cytotoxicity in human astrocytes and neurons. *Neurotox. Res.* **2009**, *16*, 77–86. [CrossRef] [PubMed]

32. Guillemin, G.J. Quinolinic acid, the inescapable neurotoxin. *FEBS J.* **2012**, *279*, 1356–1365. [CrossRef] [PubMed]

33. St'astny, F.; Lisy, V.; Mares, V.; Lisa, V.; Balcar, V.J.; Santamaria, A. Quinolinic acid induces NMDA receptor-mediated lipid peroxidation in rat brain microvessels. *Redox. Rep.* **2004**, *9*, 229–233. [CrossRef] [PubMed]

34. Marin, O. Origin of cortical interneurons: Basic concepts and clinical implications. *Rev. Neurol.* **2002**, *35*, 743–751. [PubMed]

35. Alberti, E.; Los, M.; Garcia, R.; Fraga, J.L.; Serrano, T.; Hernandez, E.; Klonisch, T.; Macias, R.; Martinez, L.; Castillo, L.; de la Cuétara, K. Prolonged survival and expression of neural markers by bone marrow-derived stem cells transplanted into brain lesions. *Med. Sci. Monit.* **2009**, *15*, BR47–BR54. [PubMed]

36. Serrano, T.; Pierozan, P.; Alberti, E.; Blanco, L.; de la Cuétara Bernal, K.; González, M.E.; Pavón, N.; Lorigados, L.; Robinson-Agramonte, M.A.; Bergado, J.A. Transplantation of mononuclear cells from bone marrow in a rat model of Huntington's disease. *Neurorestoratology* **2016**, *4*, 95–105. [CrossRef]

37. Teresa, S.-S. Havana: International Center for Neurological Restoration (CIREN). Ph.D. Thesis, Cuba University, Havana, Cuba, 2017.

behavioral sciences

MDPI

Article

Cellular Redox Imbalance and Neurochemical Effect in Cognitive-Deficient Old Rats

Maria Elena González-Fraguela [1,*], Lisette Blanco-Lezcano [2],
Caridad Ivette Fernandez-Verdecia [2], Teresa Serrano Sanchez [1],
Maria de los A. Robinson Agramonte [1] and Lidia Leonor Cardellá Rosales [3]

[1]   Immunochemical Department, International Center for Neurological Restoration, 25th Ave, Playa, 15805, PC 11300 Havana, Cuba; teresa@neuro.ciren.cu (T.S.S.); robin@neuro.ciren.cu (M.d.l.A.R.A.)
[2]   Experimental Neurophysiology Department, International Center of Neurological Restoration (CIREN) Ave. 25 No. 15805 e/158 and 160, Playa, Havana 11300, Cuba; lblanco@neuro.ciren.cu (L.B.-L.); ivettef@neuro.ciren.cu (C.I.F.-V.)
[3]   Physiologic Sciences Department, Latin American Medicine School, Carretera Panamericana, Kilómetro 3 1/2 Municipio Playa, Habana 19148, Cuba; lcardella@infomed.sld.cu
*   Correspondence: marie@neuro.ciren.cu; Tel.: +53-7-2715353

Received: 22 August 2018; Accepted: 8 October 2018; Published: 13 October 2018

Abstract: The purpose of the present study is to access the linkage between dysregulation of glutamatergic neurotransmission, oxidative metabolism, and serine signaling in age-related cognitive decline. In this work, we evaluated the effect of natural aging in rats on the cognitive abilities for hippocampal-dependent tasks. Oxidative metabolism indicators are glutathione (GSH), malondialdehyde (MDA) concentrations, and cytosolic phospholipase A_2 (PLA_2) activity. In addition, neurotransmitter amino acid (*L*-Glutamic acid, γ-aminobutyric acid (GABA), *DL*-Serine and *DL*-Aspartic acid) concentrations were studied in brain areas such as the frontal cortex (FC) and hippocampus (HPC). The spatial long-term memory revealed significant differences among experimental groups: the aged rats showed an increase in escape latency to the platform associated with a reduction of crossings and spent less time on the target quadrant than young rats. Glutathione levels decreased for analyzed brain areas linked with a significant increase in MDA concentrations and PLA_2 activity in cognitive-deficient old rats. We found glutamate levels only increased in the HPC, whereas a reduced level of serine was found in both regions of interest in cognitive-deficient old rats. We demonstrated that age-related changes in redox metabolism contributed with alterations in synaptic signaling and cognitive impairment.

Keywords: aging; learning; memory; neurotransmitter; oxidative stress; phospholipase A_2; serine

1. Introduction

The demographic changes that affect a large part of humanity are leading to an increase in life expectancy at birth, which has a high impact on societies and their economies. The increase in health costs are not at the expense of the new generation, in fact they are very reduced, but at the expense of the predominance of pathologies related to aging and other physical and intellectual disabilities associated with advanced age.

Aging is an essential factor for many diseases, specifically in neurodegenerative diseases. Oxidative stress constitutes the most important theory of ageing at the molecular level. The cognitive processes that are mediated by the hippocampus (declarative memory) and the prefrontal cortex (working memory) represent the most vulnerable areas to ageing.

This is due to brain cells that are exposed to rising levels of oxidative stress with the loss of energy homeostasis, and the increase in lipid mediators by phospholipases A_2 (PLA$_2$) activity, alterations in neurotransmitters, and signal transduction and accumulation of damage in biomolecules [1].

Oxidative stress is characterized by an imbalance in reactive oxygen species production (ROS) or impaired antioxidant protection. Both mechanisms are considered to have an important role in age-related cognitive decline. Glutathione (GSH) is the main mechanism of antioxidant defense against ROS in biological systems [2]. Glutathione concentration decreases with age in several animal models [3–5] and brain areas [6,7]. The reduction in GSH concentration causes an incomplete oxidation of the substrates into the mitochondria inducing the leakage of electrons from the electron transport chain and consequently increasing ROS generation.

A habitual target of ROS are neural membrane lipids, the enhanced phospholipid degradation induces changes in fluidity and permeability in neural membranes with the consequent alteration of ionic channels and modulation in the enzymatic activities coupled to membrane composition [8].

Previous studies have demonstrated the involvement of oxidative signaling pathways in a permanent damage of the neuronal circuitry, especially with glutamatergic neurons. Sustained activation of ionotropic glutamate receptors as the *N*-methyl-D-aspartic acid receptor (NMDAR), could affect intracellular calcium levels leading to activation of proteases, nucleases, and phospholipases [9].

The cytosolic PLA$_2$ causes the release of arachidonic acid (AA), hydrolyzing the sn-2 fatty acids of membrane phospholipids. This important polyunsaturated fatty acid (PUFA) is the precursor for the synthesis of eicosanoids and prostanoids. Some findings indicate that PLA$_2$ plays important roles in synaptic functions, long-term potentiation (LTP), learning, and memory [10]. The effect of PLA$_2$ on cognitive function can be related with the ability of AA to change neuronal metabolism and synaptic activity [11].

Glutamate, an excitatory neurotransmitter in the central nervous system (CNS), plays a recognized role in the regulation of neurogenesis, neuronal survival, synaptic plasticity, and learning and memory processes [12].

Several studies have suggested that the stimulation of PLA$_2$ is a process mediated by the glutamate receptor in neuronal cultures [10]. Likewise, D-serine is the most effective co-agonist of the glycine site of NMDAR, which is essential for the optimal function of this receptor in physiological and pathological processes [13].

Experimental data has demonstrated that D-serine treatment decreases the extent of neuron death with a neuroprotective effect against apoptosis [14]. At the present time, serine metabolism is the subject of intense investigation to confirm its role in the diagnosis and therapy of neurological disorders, as well as its therapeutic potential in preclinical models [15,16].

The objective of this study was to obtain evidence about the linkage between dysregulation of glutamatergic neurotransmission in oxidative stress conditions and the role of serine signaling in age-related cognitive decline.

2. Materials and Methods

2.1. Animal

Male Sprague–Dawley rats obtained from a national breeder institution (CENPALAB, Artemisa) were housed five per cage in a 12:12-h light-dark cycle with food and water ad libitum, under standard conditions for temperature (25 °C) and relative humidity (60%). All experimental studies were performed during light cycle (7.00–19.00 h). We used two age groups: young, ($n = 30$; 2 months old) with a body weight 272 ± 36 g (mean $\pm$ standard deviation, SD); and cognitive-deficient old rats ($n = 45$; 23 months old, body weight 520 ± 46 g).

All efforts were made to minimize the number of animals used and their suffering. All procedures concerning animal experimentation were carried out according to the ethical principles for animal research established by the Canadian Council for Animal Care [17] and the Cuban Regulations for the Use of Laboratory Animals (CENPALAB 1997) and were approved by the Ethical Committee of the International Center for Neurological Restoration.

2.2. Behavioral Test

Morris Water Maze (MWM)

The MWM tests spatial learning and memory retention in rodents and involves an acquisition phase where the animal learns the location of a "goal or escape platform" target over a period of multiple days [18]. An overhead camera and computer assisted tracking system with videomax software (Spontaneous Motor Activity Recording Tracking SMART 2.0) (PanLab, Barcelona España, Spain) was used to record the position of the rat in the maze (Scheme 1). A cued learning test was used to discriminate motoric or motivational factors. On the first day, animals received four trials in a water maze (no distal cues) with a well visible and distinguished platform located at different quadrants.

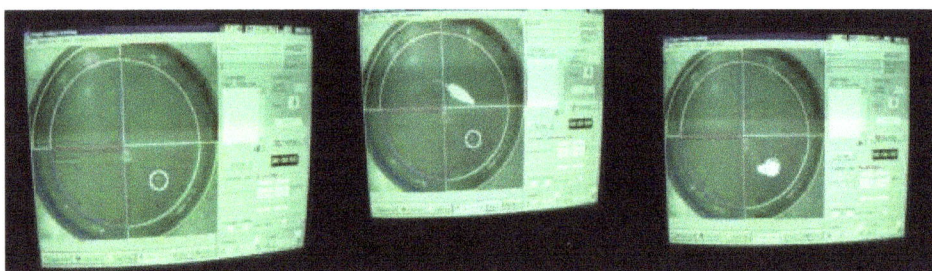

Scheme 1. The software display during data acquisition in the Morris water maze (MWM) task using the Spontaneous Motor Activity Recording Tracking (SMART) system from Panlab S.A.

One day after, a classic spatial acquisition test was introduced. Animals were trained in the MWM with non-visible platform conditions for eight trials per day over four consecutive days. The order of the start position was changed in every trial as well as the sequence of positions along the days. The platform position was kept fixed in the northeast position (NE target quadrant, Q1) throughout the training period (Scheme 2). In each trial, the rat was released from the starting position, facing the wall, and was allowed to search for a hidden escape platform for 60 s. If the animal was unable to find the platform it was gently carried onto it, and allowed a 30 s rest period sitting on the platform. Data from each four trials were grouped into one block (4 days: 32 trials: 8 blocks-B1–B8). On the fifth day (trial 33), the platform was removed and the animal was allowed to search for it for 60 s (probe trial for Q1). This trial evaluated the quality of reference memory. Immediately, the platform was relocated in the same position (Q1) to evaluate four additional trials (34–37, B9) to explore the cognitive flexibility with a previous dissociating stimulus (absence of escape platform). Two days later, the rat's long-term reference memory was evaluated in a single block of four trials (B10) using the above described paradigm (Q1 at NE position).

Aged rats are classified according to the criteria of 2 or 2.5 standard deviations above the average of the young controls latency [19–21].

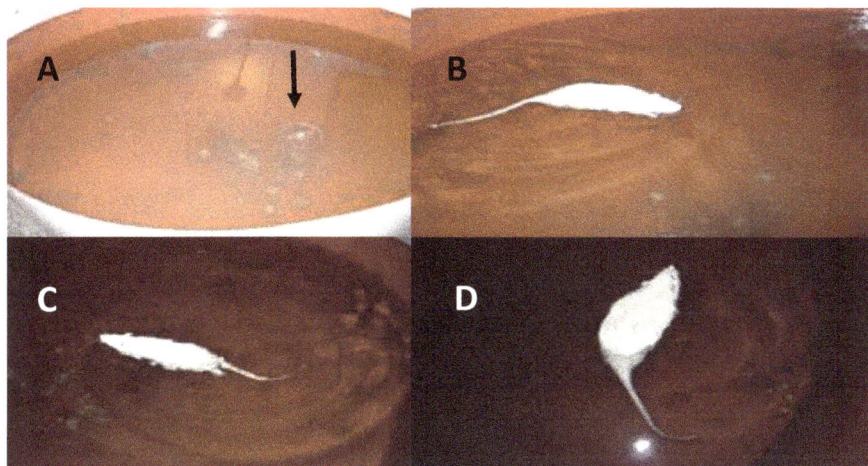

Scheme 2. (**A**) The tank before the task. The arrow shows the location of the platform. (**B–D**) The different moments in which the rat is searching for the platform in the maze and reaching it.

2.3. Sedation and Sample Collection

For the studies that required sedation, an anesthetic mixture of 4 mL of ketalar (ketamine 50 mg/mL, IMEFA, Havana, Cuba), 2 mL of valium (diazepam 5 mg/mL, IMEFA, Havana, Cuba), 1 mL of atropine (atropine, 0.5 mg/mL, IMEFA, Havana, Cuba), and 16 mL of 0.9% sodium chloride solution (0.9% NaCl, IMEFA, Havana, Cuba) was used. The dose regimen was 1 mL of the mixture/100 g of body weight, and immediately the rats were sacrificed by decapitation. Their brains were extracted and rinsed in cold saline; the total frontal cortex (FC) and hippocampus (HPC) were bilaterally dissected. The tissues were frozen in liquid nitrogen, weighed, and stored at −80 °C until the biochemical analysis.

2.4. Biochemical Analysis

2.4.1. Oxidative Stress Markers

Glutathione GSH Quantification

Tissue samples were homogenized in a glass-Teflon potter containing 5% 5-sulfosalicylic acid (1:15, w/v) and the protein-free supernatant was isolated by centrifugation at 8160× g during 10 min. Total GSH was quantified by Tietze's recycling assay as described by Azbill et al. [22]. Aliquots of the supernatants were incubated for 25 min at 37 °C in a medium containing 0.21 mM NADPH, 0.6 mM DTNB (5,5'-ditiobis(2-nitrobenzoico), 6.3 mM EDTA (ethylene diamine tetraacetic acid), and 143 mM sodium phosphate pH 7.5. Following the addition of 0.5 U GRD (glutathione reductasa), the absorbance increase at 412 nm due to the formation of 5-thio-nitrobenzoate was recorded. The concentration values were calculated from a GSH standard curve.

Malondialdehyde (MDA) Quantification

Tissues samples were homogenized in 1 M Tris/0.25 M sucrose buffer (pH 7.4) at a tissue/buffer volume ratio of 1/5. The homogenates were centrifuged at 14,000 rpm for 15 min. Two-hundred microliters of the sample were vortexed with 400 µL of a 0.67% thiobarbituric acid solution in 0.2 M HCl, then incubated for 15 min in a water bath at 100 °C and centrifuged at 2448× g for 10 min at room temperature. The resulting supernatant was used for measuring absorbance at 535 nm. The concentration values were calculated from an MDA standard curve [23].

Assay for Calcium-Dependent Cytosolic Phospholipase A2 (PLA2) Activity

PLA$_2$ enzymatic activity was measured using the protocol recommended by the manufacturer (Cayman Chemical Company, Ann Arbor, MI, USA). Briefly, 10 µL sample were incubated with 1.5 mM arachidonoyl thio-phosphotidylcholine (a PLA$_2$ substrate) for 1 h at room temperature. Enzyme catalysis was stopped by the addition of beta dystrobrevin/EGTA (Ethylene glycol-bis(2-aminoethylether)-*N*,*N*,*N'*,*N'*-tetraacetic acid), and the optical density of the sample was measured at 405 nm. PLA$_2$ activity was calculated in accordance to kit instruction and PLA$_2$ activity was expressed in U/mg [24].

2.4.2. Amino Acid Analysis

Sample Preparation and Derivatization

The brain tissue for amino acid analysis was homogenized in ice-cold 0.2 M perchloric acid including cysteine (4 mL per g of tissue) and sonicated for 5 min. The brain homogenate was centrifuged at 17,000 rpm for 4 min at 4 °C and the supernatant was diluted (1/50), aliquoted, and stored at −20 °C. 20 µL of standard (*L*-Glutamic acid, γ-aminobutyric acid (GABA), *DL*-Serine and *DL*-Aspartate) or sample was added with 20 µL of the derivatization reagent (25 mg OPA (Ophthaldialdehyde) of diluted in 1500 µL of methanol, 100 µL of borate buffer 0.1 M (pH9), and 30 µL of β-mercaptoethanol), the derivatization mixture was vortexed for 1 min at room temperature, After that, 5 µL of 5% acetic acid were added to stop of reaction, then, 20 µL of this final solution was injected into the HPLC system.

Instrumentation and Chromatographic Conditions

For chromatographic analysis, an HPLC system (Agilent 1200 Technologies) (Santa Clara, CA, USA) was used. Chromatographic separation was achieved on a reversed phase column (HR-80, ESA), using an isocratic chromatographic pump (Agilent 1200 series), and detected with a fluorescence detector (Agilent G1321A, 1100 series) set at $\lambda_{excitation}$ 340 nm and $\lambda_{emission}$ 460 nm. The column temperature was maintained at 25 °C and the mobile phase was pumped at a flow rate of 1.0 mLmin^{-1}. The chromatographic data were recorded using the Agilent ChemStation B.04.02SP1 software. The mobile phase was composed of 0.1 M NaH$_2$PO$_4$/25% methanol.

2.5. Data Processing and Statistical Analysis

The values were expressed as mean ± standard error mean. Normal distribution and homogeneity of variance of the data were tested by the Kolmogorov–Smirnov and Levene tests, respectively. The comparison between experimental groups of the biochemical indicators was performed by a one-way analysis of variance (ANOVA) and the behavioral variables in the MWM were analyzed by analysis of variance for repeated measurements (ANOVA). For post hoc comparisons was used the Tukey honest significant differences test (HSD) and nonparametric comparison with Mann–Whitney U test. Significant differences were considered only if $p < 0.05$.

3. Results

3.1. Behavioural Studies

Morris Water Maze

The post-hoc analysis on spatial long-term memory revealed significant differences between experimental groups for escape latency to platform ($F_{(1, 28)}$ = 12.39; $p < 0.001$) related with an increase in aged rats that demonstrated severe spatial memory impairment in this group (Figure 1). Likewise, regarding the number of crossings, the aged rats exhibited a poor performance in the acquisition of the navigation search strategy during the test. The Tukey honest significance differences test, showed a reduction of crossings and time spent on the target quadrant in older animals than younger rats ($F_{(1, 28)}$ = 8.91; $p < 0.001$) (Figure 2, Table 1).

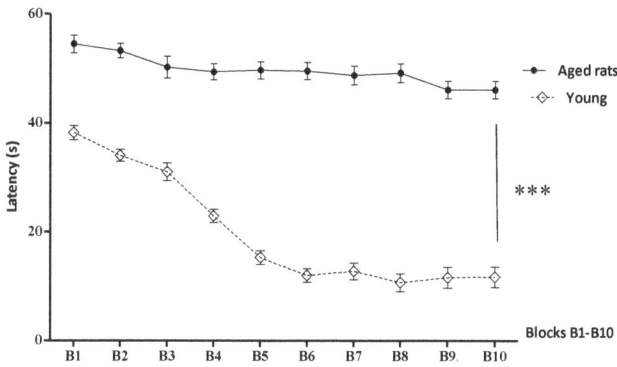

Figure 1. Latency in the MWM (mean ± SD mean are represented) to find the hidden platform during the training sessions grouped into the blocks (B). Classical training (B1–B9) and long-term contextual memory (B10). *** Significant differences between experimental groups: aged rats ($n = 45$) and young ($n = 30$) ($p < 0.001$).

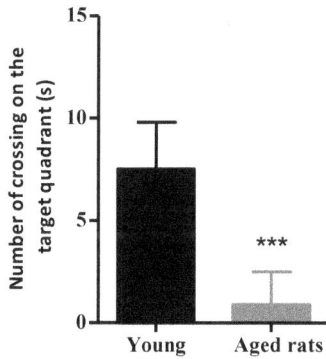

Figure 2. Number of crossings on the target quadrant (mean ± SD mean are represented) *** Significant differences between experimental groups: aged rats ($n = 45$) and young ($n = 30$) ($p < 0.001$).

Table 1. Data (mean ± SD mean) from four trials grouped in blocks (B1–B10) for the *latency* in the MWM and the number of crossings in the target quadrant between experimental groups.

Blocks (B)	Aged Rats (Mean ± SD Mean)	Young (Mean ± SD Mean)
B1	54.5 ± 1.62	38.20 ± 1.32
B2	53.30 ± 1.33	34.02 ± 1.087
B3	50.26 ± 1.97	30.99 ± 1.61
B4	49.41 ± 1.48	22.96 ±1.20
B5	49.7 ± 1.56	15.30 ± 1.27
B6	49..59 ± 1.54	11.96 ± 1.26
B7	48.8 ± 1.70	12.75 ± 1.57
B8	48.2 ± 1.73	10.66 ± 1.59
B9	46.12 ± 1.60	11.58 ± 1.94
B10	46.06 ± 1.7	11.70 ± 1.92
Number of Crossings	0.9 ± 1.6	7.5 ± 2.3

3.2. Biochemical Indicators

The comparison between experimental groups with regard to GSH content showed significant differences in examined brain areas in cognitive-deficient old rats (FC ($Z = 5.41$; $p < 0.001$); HPC ($F_{(1,38)}$ = 288.6; $p < 0.001$)). The level of GSH dropped by 82% in FC and 80% in HPC in respect to the control groups. These results are associated with a significant increase in MDA concentrations for aged rats only for hippocampus, a highly sensitive region to oxidative stress. (HPC ($F_{(1,38)} = 837.9$; $p < 0.001$)).

Similarly, a considerable increase was observed in PLA$_2$ activity for analyzed brain areas. The enzyme activity revealed significant differences among experimental groups (FC ($F_{(1,38)} = 29.15$; $p < 0.001$); HPC ($F_{(1,38)} = 70.96$, $p < 0.001$)) in correspondence with the reduction of GSH content in the studied brain structures GSH: Glutathione, MDA: Malondialdehyde, PLA$_2$: Phospholipase A$_2$, GABA: γ-aminobutyric acid. (Figure 3, Table 2).

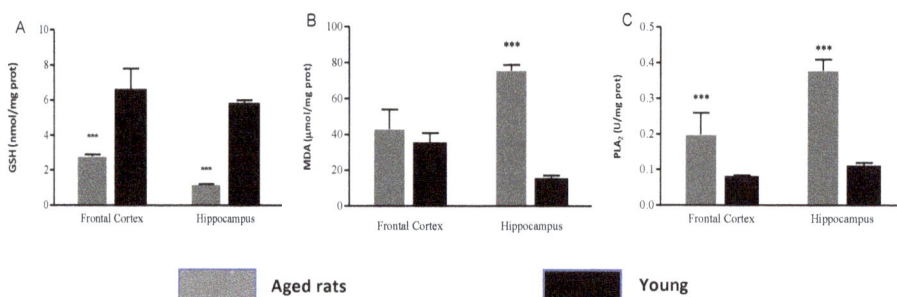

Figure 3. Effect of aging on oxidative metabolism indicators in the examined brain areas. (**A**) Glutathione (GSH), (**B**) Malondialdehyde (MDA), (**C**) Phospholipase A2 (PLA2). *** Corresponds to significant differences between experimental groups ($p < 0.001$).

Table 2. Data (mean ± SD mean) corresponding to biochemical indicators in examined brain areas between experimental groups.

Biochemical Indicators	Frontal Cortex Aged Rats	Frontal Cortex Young Rats	Hippocampus Aged Rats	Hippocampus Young Rats
GSH	1.3 ± 0.1	5.45 ± 1.2	0.76 ± 0.22	5.15 ± 0.2
MDA	43.42 ± 10.6	34.93 ± 5.95	80.31 ± 3.01	15.84 ± 2.02
PLA$_2$	0.2 ± 0.06	0.08 ± 0.0	0.35 ± 0.03	0.11 ± 0.01
L Glutamate	5.29 ± 0.8	5.17 ± 1.10	10.24 ± 1.2	6.89 ± 1.2
GABA	0.88 ± 0.03	0.94 ± 0.02	1.11 ± 0.03	1.17 ±0.03
DL Serine	4.85 ± 0.8	6.07 ± 0.55	2.13 ± 1.02	5.52 ± 1.2
DL Aspartate	5.35 ± 0.72	5.4 ± 0.55	5.02 ± 1.03	5.16 ± 1.2

Amino Acid Concentrations

The levels of GABA were unchanged compared with controls in brain regions investigated in aged rats (FC ($F_{(1,38)} = 0.17$; $p > 0.05$) and HPC ($F_{(1,38)} = 0.39$; $p > 0.05$)). Furthermore, the aspartate concentrations were not significantly different between groups (FC ($F_{(1,38)} = 0.18$; $p > 0.05$) and HPC ($F_{(1,38)} = 0.70$; $p > 0.05$)). In contrast, the levels of glutamate only increased in HPC in aged rats (HPC ($F_{(1,38)} = 93.32$; $p < 0.001$), whereas a reduced level of serine was found in the regions of interest in cognitive-deficient old rats (FC ($F_{(1,38)} = 58.3$; $p < 0.001$) and HPC ($F_{(1,38)} = 365.1$; $p < 0.001$)) (Figure 4, Table 2).

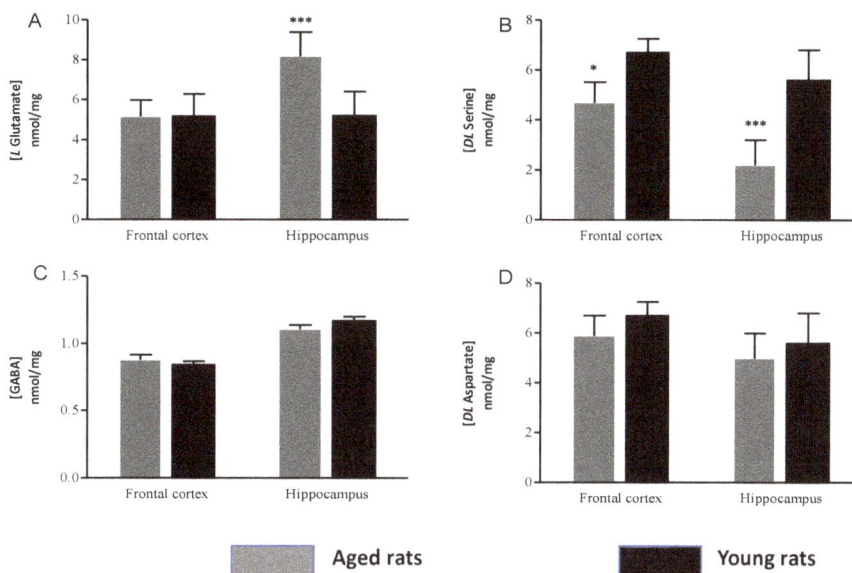

Figure 4. Effect of aging on amino acid concentration in the examined brain areas. (**A**) *L* Glutamate, (**B**) *DL* Serine, (**C**) GABA, (**D**) *DL* Aspartate. *** ($p < 0.001$); * ($p < 0.05$) corresponds to significant differences between experimental groups.

4. Discussion

Some aged animals undergo severe cognitive deficiencies, whereas others preserve their normal cognitive abilities. In particular, spatial memory appears to be significantly affected [25]. The behavioral results allowed us to select aged rats that demonstrated a deficient execution in the acquisition of a successful navigation strategy in the MWM. The metabolic changes originated by GSH depletion in HPC caused increased swim times in the MWM test associated with a reduction of time spent on the target quadrant, confirming a failure in the memory trace formation, and therefore cognitive dysfunction.

We observed a significant GSH depletion in cognitive-deficient old rats, as much in FC as in HPC. Besides its roles in maintenance of oxidant homeostasis within the cell, GSH may serve as a substrate and an allosteric modulator of eicosanoid biosynthesis, in the regulation of NMDAR activity and activation of transcription factors [26]. We found a reduction of more than 70% in the GSH content in cognitive-deficient old rats compared to the young rats. The majority of authors have reported a decreases of tripeptide in brain tissue of around 50% in previous rodent studies [27,28].

The strong GSH decrease observed in our work generates hydrogen peroxide (H_2O_2) and hydroxyl radical (OH) that induce lipid peroxidation in the membrane of neurons and glial cells and damage the (Ca^{2+} and Na^+/K^+) ATPases and glucose transporters [29]. In connection with this result, we observed a significant increase in MDA concentration and PLA$_2$ activity in aged rats for examined brain areas. The loss of the energy metabolism homeostasis, redox imbalance, and subsequent damage by accumulation of MDA could originate the functional impairment in aged rats.

Due to its numerous functions for maintenance of membrane phospholipid balance and for originating several lipid mediators, PLA$_2$ activation has been implicated in pathological conditions in the CNS for neurodegenerative diseases among others [8,30–32].

H_2O_2 accumulation induces the redox ambience appropriate for activating the PLA$_2$ pathway by means of extracellular signal-regulated kinase (ERK) and interleukin 1b [33,34]. In this context,

PLA$_2$ activity increases the release of AA from the plasma membrane, which directly produce 4-hydroxynonenal and MDA.

Arachidonic acid AA undergoes oxidative modifications and increases the mitochondrial ROS levels in the hippocampus and cerebral cortex [35]. As such, the AA and its metabolites are permeable to the membrane and could affect the function of neighboring neurons with disruption of monolayer integrity and cell death [36].

The results obtained from the cognitive-deficient old rats revealed high PLA$_2$ enzymatic activity and consequently increased AA generation. It is important to point out that high concentrations of AA cause uncoupling oxidative phosphorylation with a detrimental effect on the ATP-producing capacity of mitochondria, which results in mitochondrial dysfunction and an additional increase in the production of ROS.

These experimental data are in line with clinical studies. Jiang R et al. [37] reported that higher PLA$_2$ enzymatic activity is associated with increased prevalence of cognitive impairment in the Chinese population. On the other hand, there is evidence that the contribution of PLA$_2$ activity in regulating neuronal excitatory functions, in cultured primary cortical neurons, trigger NMDAR receptors, which has been demonstrated to activate PLA$_2$ and AA release [38]. In turn, PLA$_2$ activity and AA are involved in receptor signaling which is linked with oxidative events which underline important roles in synaptic signaling, LTP, learning and memory [39,40].

In consonance with these studies, the present findings evidence an increase in glutamate concentrations in the HPC of aged rats which is closely associated with the elevation of PLA$_2$ activity in this region. The LTP dependence of NMDAR activation has been demonstrated in several brain areas, including hippocampal subfields like the dentate gyrus and the CA1 [41].

It is well established now that excessive levels of glutamate can produce toxic effects due to overstimulation of NMDAR, increasing the intracellular Ca^{2+} and activate Ca^{2+}-dependent enzymes including PLA$_2$. This cascade of events culminates with a Ca^{2+} dysregulation in neurons and astrocytes, which initiates cellular death by a process known as excitotoxicity. This process implicates the activation of proteases, the generation of radical species, and mitochondrial Ca^{2+} overload that affects critical biological molecules including lipids, proteins, and DNA.

However, on the other hand, one of the deleterious consequences of GSH metabolism dysfunction is the NMDAR hypoactivity [42]. The GSH insufficiency in aged rats could induce the formation of disulfide bonds between pairs of cysteine residues on the extracellular portion of the NMDAR generating a selective reduction of NMDAR currents [42], which contrast with the possible excitotoxic effect of NMDAR activation in old rats due to high levels of glutamate.

Previous studies with NMDAR blockade suggested that altered redox states of these extracellular cysteine residues could mediate the decline in NMDAR function during aging [43]. The application of dithiothreitol, a specific reducing agent, increased the NMDAR component and the magnitude of LTP in hippocampal slices from old rodents [43].

The intracellular redox status observed in the aged rats can directly contribute to synaptic inefficiency of NMDAR, caused by changes in neurotransmitter systems in vulnerable brain regions. Liu et al. [44] have reported that H_2O_2 accumulation is associated with the downregulation of the GluN2 subunits of the NMDAR in adult rat hippocampi. Otherwise, recent modeling work has shown that reduced NMDAR signaling into interneurons produced alterations in spatial representations that affected working memory execution [45]. Therefore, the suppression of NMDAR responses could induce changes in memory mechanisms that impaired the cognitive processes in GSH-depleted aged rats.

This previous hypothesis is supported by a reduced level of serine in the brain regions investigated in aged rats. D-serine, an enantiomer of serine, has a crucial characteristic of binding to NMDAR [46]. This amino acid binds to the glycine-site of the GluN1 subunit of NMDAR so that glutamate can link to GluN2 subunits [47].

We found a reduction of around 60% in serine levels in HPC and 30% in FC in cognitive-deficient old rats with regard to younger counterparts. The bond of D-serine is critical to synaptic NMDAR function [48] due to D-serine having three additional hydrogen bonds to the receptor, which may activate the NMDAR more efficiently than glycine [49]. This finding increases the possibility that alterations in serine metabolism could be a contributor to the dysfunctions of memory in the aged rats.

D-serine is converted from L-serine by serine racemase, an enzyme dependent of pyridoxal phosphate, mostly expressed in excitatory neurons in the CNS of rodents [50]. D-serine is especially increased in brain areas involved in cognitive processes, as in the cerebral cortex, hippocampus, and basal ganglia [49].

The pathway responsible for the degradation of D-serine in the CNS is the D-amino acid oxidase (DAAO) [51]. Nevertheless, the reduction in serine concentration could involve an exceptional property of serine racemase which is able to catalyze the degradation of D-serine and L-serine to pyruvate and ammonia [13].

The possibility of serine racemase to enhance pyruvate levels from metabolic availability of serine may represent an important mechanism to obtain energy for mitochondrial oxidative phosphorylation and synthesis of ATP. This alternative of the serine metabolic pathway explains how the increased ROS production secondary to GSH depletion observed in aged rats reduces the synaptic levels of D-serine and exacerbates the NMDAR dysfunction in increased energy demand conditions.

These results are coincidental with reports from others authors who have shown that alterations of LTP expression are associated with a deficiency of NMDAR [52–54] and is recovered in aged rats and senescence-accelerated mice by D-serine treatment [54].

5. Conclusions

These findings suggest that age-related changes in the cellular redox metabolism contributes to memory dysfunctions and cognitive decline. The energy demands of synaptic activity lead to increased ROS production as a by-product of ATP synthesis with elevated utilization of a glutathione-based antioxidant system. Glutathione depletion induces an acute increase in the oxidative stress conditions and may generate NMDAR hypofunction and alterations in synaptic transmission. In the same way, the reduction of serine levels in aged rats, as the most effective co-agonist, could lead to a marked impairment of NMDAR function. The results discussed here support the hypothesis of redox homeostasis as a crucial element that regulates brain function and it will be necessary to further research the molecular mechanism of observed results.

Author Contributions: This work was carried out in collaboration between all authors. M.E.G.-F. was the principal investigator, designed the study, conducted the data analysis and wrote of the paper. L.B.-L. and C.I.F.-V. supervised the statistical analysis and assisted with interpreting the results and revision of the manuscript. T.S.S. and M.d.l.A.R.A. participated in the analysis of the data. L.L.C.R. participated in the analysis of the data and revision of the manuscript. All authors read and approved the final manuscript.

Funding: This study was supported financially by the CIREN Scientific Council.

Conflicts of Interest: The authors declare no conflict of interest.

References

1. Zhang, Y.; Li, P.; Feng, J.; Minghu, W. Dysfunction of NMDA receptors in Alzheimer's disease. *Neurol. Sci.* **2016**, *37*, 1039–1047. [CrossRef] [PubMed]
2. Mari, M.; Morales, A.; Colell, A.; Garcia-Ruiz, C.; Fernandez-Checa, J. Mitochondrial Glutathione, a Key Survival Antioxidant. *ARS* **2009**, *11*, 2685–2700. [CrossRef] [PubMed]
3. Bergado, J.A.; Almaguer, W.; Rojas, J.; Capdevila, V.; Frey, J.U. Spatial and emotional memory in aged rats: A behavioral-statistical analysis. *Neuroscience* **2011**, *172*, 256–269. [CrossRef] [PubMed]

4. Krämer, T.; Grob, T.; Menzel, L.; Hirnet, T.; Griemert, E.; Radyushkin, K.; Thal, S.C.; Methner, A.; Schaefer, M.K.E. Dimethyl fumarate treatment after traumatic brain injury prevents depletion of antioxidative brain glutathione and confers neuroprotection. *J. Neurochem.* **2017**, *143* (Suppl. 5), 523–533. [CrossRef] [PubMed]
5. Gemelli, T.; de Andrade, R.B.; Rojas, D.B.; Zanatta, Â.; Schirmbeck, G.H.; Funchal, C.; Wajner, M.; Dutra-Filho, C.S.; Wannmacher, C.M.D. Chronic Exposure to β-Alanine Generates Oxidative Stress and Alters Energy Metabolism in Cerebral Cortex and Cerebellum of Wistar Rats. *Mol. Neurobiol.* **2018**, *55* (Suppl. 6), 5101–5110. [CrossRef] [PubMed]
6. Cruz-Aguado, R.; Almaguer-Melian, W.; Diaz, C.M.; Lorigados, L.; Bergado, J. Behavioral and biochemical effects of glutathione depletion in the rat brain. *Brain Res. Bull.* **2001**, *55* (Suppl. 3), 327–333. [CrossRef]
7. Bonasera, S.; Arikkath, J.; Boska, M.D.; Chaudoin, T.R.; De Korver, N.W.; Goulding, E.H.; Traci, A.; Hoke, T.A.; Mojtahedzedah, V.; Reyelts, C.D.; et al. Age-related changes in cerebellar and hypothalamic function accompany non-microglial immune gene expression, altered synapse organization, and excitatory amino acid neurotransmission deficits. *Aging* **2016**, *8*, 2153–2164. [CrossRef] [PubMed]
8. Chauhan, V.; Chauhan, A. Abnormalities in membrane lipids, membrane-associated proteins, and signal transduction in Autism. In *Austim Oxidative Stress, Inflammation and Immune Abnormalities*; Chauhan, A., Ed.; CRC Press: Boca Raton, FL, USA, 2010; pp. 177–207.
9. Paoletti, P.; Bellone, C.; Zhou, Q. NMDA receptor subunit diversity: Impact on receptor properties, synaptic plasticity and disease. *Nat. Neurosci.* **2013**, *14*, 383–400. [CrossRef] [PubMed]
10. Basselin, M.; Chang, L.; Chen, M.; Bell, J.M.; Rapoport, S. Chronic Administration of Valproic Acid Reduces Brain NMDA Signaling via Arachidonic Acid in Unanesthetized Rats. *Neurochem. Res.* **2008**, *33*, 2229–2240. [CrossRef] [PubMed]
11. Ng, C.Y.; Kannan, S.; Chen, Y.J.; Tan, F.C.; Ong, W.Y.; Go, M.L.; Verma, C.S.; Low, C.M.; Lam, Y. A New Generation of Arachidonic Acid Analogues as Potential Neurological Agent Targeting Cytosolic Phospholipase A2. *Sci. Rep.* **2017**, *7*, 13683. [CrossRef] [PubMed]
12. Wang, R.; Reddy, P.H. Role of glutamate and NMDA receptors in Alzheimer's disease. *J. Alzheimers Dis.* **2017**, *57* (Suppl. 4), 1041–1048. [CrossRef] [PubMed]
13. Billard, J.M. D-serine signalling as a prominent determinant of neuronal-glial dialogue in the healthy and diseased brain. *J. Cell. Mol. Med.* **2008**, *12*, 1872–1884. [CrossRef] [PubMed]
14. Sasabe, J.; Miyoshi, Y.; Suzuki, M.; Mita, M.; Konno, R.; Matsuoka, M.; Hamase, K.; Aiso, S. D-amino acid oxidase controls motoneuron degeneration through D-serine. *Proc. Natl. Acad. Sci. USA* **2012**, *109*, 627–632. [CrossRef] [PubMed]
15. Nuechterlein, K.H.; Subotnik, K.L.; Green, M.F.; Ventura, J.; Asarnow, R.F.; Gitlin, M.J.; Yee, C.M.; Gretchen-Doorly, D.; Mintz, J. Neurocognitive predictors of work outcome in recent-onset schizophrenia. *Schizophr. Bull.* **2011**, *37*, 33–40. [CrossRef] [PubMed]
16. Fujita, Y.; Ishima, T.; Hashimoto, K. Supplementation with D-serine prevents the onset of cognitive deficits in adult offspring after maternal immune activation. *Sci. Rep.* **2016**, *6*, 37261. [CrossRef] [PubMed]
17. Olferd, E.; Cross, B.; McWillian, D.; McWillian, A. *Guidelines for the Use of Animal in Neuroscience Research*; Canadian Council on Care (CCAC), Bradda Printing Services Inc.: Ottawa, ON, Canada, 1997; pp. 163–165.
18. Van Praag, H.; Shubert, T.; Zhao, C.; Gage, F.H. Exercise enhances learning and hippocampal neurogenesis in aged mice. *J. Neurosci.* **2005**, *25* (Suppl. 38), 8680–8685. [CrossRef] [PubMed]
19. Almaguer-Melian, W.; Cruz-Aguado, R.; Riva, C.L.; Kendrick, K.M.; Frey, J.U.; Bergado, J. Effect of LTP-reinforcing paradigms on neurotransmitter release in the dentate gyrus of young and aged rats. *Biochem. Biophys. Res. Commun.* **2005**, *327*, 877–883. [CrossRef] [PubMed]
20. Tombaugh, G.C.; Rowe, W.B.; Chow, A.R.; Michael, T.H.; Rose, G.M. Theta-frequency synaptic potentiation in CA1 in vitro distinguishes cognitively impaired from unimpaired aged Fischer 344 rats. *J. Neurosci.* **2002**, *22*, 9932–9940. [CrossRef] [PubMed]
21. Rowe, W.B.; Spreekmeester, E.; Meaney, M.J.; Quirion, R.; Rochford, J. Reactivity to novelty in cognitively-impaired and cognitively unimpaired aged rats and young rats. *Neuroscience* **1998**, *83*, 669–680. [CrossRef]
22. Tietze, F. Enzymic method for quantitative determination of nanogram amounts of total and oxidized glutathione: Applications to mammalian blood and other tissues. *Anal. Biochem.* **1969**, *27*, 502–522. [CrossRef]
23. Buege, J.A.; Aust, S.D. Microsomal lipid peroxidation. *Methods Enzymol.* **1978**, *52*, 302–310. [PubMed]

24. Lucas, K.K.; Dennis, E.A. Distinguishing phospholipase A2 types in biological samples by employing group-specific assays in the presence of inhibitors. *Prostag. Other Lipid Mediat.* **2005**, *77*, 235–248. [CrossRef] [PubMed]
25. Shukitt-Hale, B.; Casadesus, G.; Cantuti-Castelvetri, I.; Joseph, J.A. Effect of age on object exploration, habituation, and response to spatial and nonspatial change. *Behav. Neurosci.* **2001**, *115*, 1059–1064. [CrossRef] [PubMed]
26. Johnson, W.; Wilson-Delfosse, A.L.; Mieyal, J.J. Dysregulation of Glutathione Homeostasis in Neurodegenerative Diseases. *Nutrients* **2012**, *4*, 1399–1440. [CrossRef] [PubMed]
27. Allam, F.; Dao, T.; Gaurav, C.; Bohar, R.; Farzan, J. Grape Powder supplementation prevents oxidative stress-induced anxiety-like behaviour, memory impairment and high blood pressure in rats. *J. Nutr.* **2013**, *8*, 835–841. [CrossRef] [PubMed]
28. Blanco-Lezcano, L.; Jimenez-Martin, J.; Díaz-Hung, M.L.; Alberti-Amador, E.; Wong-Guerra, M.; González-Fraguela, M.E.; Estupiñán-Díaz, B.; Serrano-Sánchez, T.; Francis-Turner, L.; Delgado-Ocaña, S.; et al. Motor dysfunction and alterations in glutathione concentration, cholinesterase activity, and BDNF expression in substantia nigra pars compacta in rats with pedunculopontine lesion. *Neuroscience* **2017**, *348*, 83–97. [CrossRef] [PubMed]
29. Mattson, M.P. ER calcium and Alzheimer's disease: In a state of flux. *Sci. Signal.* **2010**, *3*, pe10. [CrossRef] [PubMed]
30. Rao, J.S.; Kellom, M.; Reese, E.A.; Rapoport, S.I.; Kim, H.-W. Dysregulated glutamate and dopamine transporters in postmortem frontal cortex from bipolar and schizophrenic patients. *J. Affect. Disord.* **2012**, *136*, 63–71. [CrossRef] [PubMed]
31. Chalimoniuk, M.; Stolecka, A.; Zieminska, E.; Stepien, A.; Langfort, J.; Strosznajder, J.B. Involvement of multiple protein kinases in cPLA2 phosphorylation, arachidonic acid release, and cell death in vivo and in vitro models of 1-methyl-4-phenylpyridinium-induced parkinsonism—The possible key role of PKG. *J. Neurochem.* **2009**, *110*, 307–317. [CrossRef] [PubMed]
32. Last, V.; Williams, A.; Werling, D. Inhibition of cytosolic phospholipase A2 prevents prion peptide-induced neuronal damage and co-localisation with beta iii tubulin. *BMC Neurosci.* **2012**, *13*, 106. [CrossRef] [PubMed]
33. Beckhauser, T.F.; Francis-Oliveira, J.; De Pasquale, R. Reactive Oxygen Species: Physiological and Physiopathological Effects on Synaptic Plasticity. *J. Exp. Neurosci.* **2016**, *10* (Suppl. 1), 23–48. [CrossRef] [PubMed]
34. Serrano, F.; Chang, A.; Hernandez, C.; Pautler, R.G.; Sweatt, J.D.; Klann, E. NADPH oxidase mediates beta-amyloid peptide-induced activation of ERK in hippocampal organotypic cultures. *Mol. Brain* **2009**, *2*, 31. [CrossRef] [PubMed]
35. Adibhatla, R.M.; Hatcher, J.F.; Dempsey, R.J. Phospholipase A2, hydroxyl radicals, and lipid peroxidation in transient cerebral ischemia. *ARS* **2003**, *5* (Suppl. 5), 647–654. [CrossRef] [PubMed]
36. Cocco, T.; Di Paola, M.; Papa, S.; Lorusso, M. Arachidonic acid interaction with the mitochondrial electron transport chain promotes reactive oxygen species generation. *Free Radic. Biol. Med.* **1999**, *27* (Suppl. 1–2), 1–2). [CrossRef]
37. Jiang, R.; Chen, S.; Shen, Y.; Wu, J.; Chen, S.; Wang, A.; Wu, S.; Zhao, X. Higher Levels of Lipoprotein Associated Phospholipase A2 is associated with Increased Prevalence of Cognitive Impairment: The APAC Study. *Sci. Rep.* **2016**, *6*, 33073. [CrossRef] [PubMed]
38. Shelat, P.B.; Chalimoniuk, M.; Wang, J.H.; Strosznajder, J.B.; Lee, J.C.; Sun, A.Y.; Simonyi, A.; Sun, G.Y. Amyloid beta peptide and NMDA induce ROS from NADPH oxidase and AA release from cytosolic phospholipase A2 in cortical neurons. *J. Neurochem.* **2008**, *106* (Suppl. 1), 45–55. [CrossRef] [PubMed]
39. Snyder, E.M.; Nong, Y.; Almeida, C.G.; Paul, S.; Moran, T.; Choi, E.Y.; Nairn, A.C.; Salter, M.W.; Lombroso, P.J.; Gouras, G.K.; et al. Regulation of NMDA receptor trafficking by amyloid-beta. *Nat. Neurosci.* **2005**, *8*, 1051–1058. [CrossRef] [PubMed]
40. Calon, F.; Lim, G.P.; Morihara, T.; Yang, F.; Ubeda, O.; Salem, N., Jr.; Frautschy, S.A.; Cole, G.M. Dietary n-3 polyunsaturated fatty acid depletion activates caspases and decreases NMDA receptors in the brain of a transgenic mouse model of Alzheimer's disease. *Eur. J. Neurosci.* **2005**, *22*, 617–626. [CrossRef] [PubMed]
41. Bergado, A.; Lucas, M.; Richter-Levin, G. Emotional tagging—A simple hypothesis in a complex reality. *Prog. Neurobiol.* **2011**, *94* (Suppl. 1), 64–76. [CrossRef] [PubMed]

42. Baxter, P.S.; Bell, K.F.S.; Hasel, P.; Kaindl, A.M.; Fricker, M.; Thomson, D.; Cregan, S.P.; Gillingwater, T.H.; Hardingham, G.H. Synaptic NMDA receptor activity is coupled to the transcriptional control of the glutathione system. *Nat. Commun.* **2015**, *6*, 6761. [CrossRef] [PubMed]

43. Foster, C.F. Dissecting the age-related decline on spatial learning and memory tasks in rodent models: *N*-methyl-D-aspartate receptors and voltage-dependent Ca^{2+} channels in senescent synaptic plasticity. *Prog. Neurobiol.* **2012**, *96* (Suppl. 3), 283–303. [CrossRef] [PubMed]

44. Liu, C.; Liu, Y.; Yang, Z. Myocardial infarction induces cognitive impairment by increasing the production of hydrogen peroxide in adult rat hippocampus. *Neurosci. Lett.* **2014**, *560*, 112–116. [CrossRef] [PubMed]

45. Murray, J.D.; Antigenic, A.; Gancsos, M.; Ichinose, M.; Corlett, P.R.; Krystal, J.H.; Wang, X.J. Linking microcircuit dysfunction to cognitive impairment: Effects of disinhibition associated with schizophrenia in a cortical working memory model. *Cereb. Cortex* **2014**, *24*, 859–872. [CrossRef] [PubMed]

46. Abe, T.; Suzuki, M.; Sasabe, J.; Takahashi, S.; Unekawa, M.; Mashima, K.; Iizumi, T.; Hamase, K.; Konno, R.; Aiso, S.; et al. Cellular origin and regulation of D- and L serine in in vitro and in vivo models of cerebral ischemia. *J. Cereb. Blood Flow Metab.* **2014**, *34*, 1928–1935. [CrossRef] [PubMed]

47. Balu, D.T.; Coyle, J.T. The NMDA receptor 'glycine modulatory site' in schizophrenia: D-serine, glycine, and beyond. *Curr. Opin. Pharmacol.* **2015**, *20*, 109–115. [CrossRef] [PubMed]

48. Benneyworth, M.A.; Li, Y.; Basu, A.C.; Bolshakov, V.Y.; Coyle, J.T. Cell selective conditional null mutations of serine racemase demonstrate a predominate localization in cortical glutamatergic neurons. *Cell. Mol. Neurobiol.* **2012**, *32*, 613–624. [CrossRef] [PubMed]

49. Suzuki, M.; Imanishi, N.; Mita, M.; Hamase, K.; Aiso, S.; Sasabe, J. Heterogeneity of D-Serine Distribution in the Human Central Nervous System. *ASN Neuro* **2017**, 1–10. [CrossRef] [PubMed]

50. Wolosker, H.; Balu, D.T.; Coyle, J.T. The Rise and Fall of the d-Serine-Mediated Gliotransmission Hypothesis. *Trends Neurosc.* **2016**, *39*, 712–721. [CrossRef] [PubMed]

51. Vanoni, M.A.; Cosma, A.; Mazzeo, D.; Mattevi, A.; Todone, F.; Curti, B. Limited proteolysis and X-ray crystallography reveal the origin of substrate specificity and of the rate-limiting product release during oxidation of D-amino acids catalyzed by mammalian D-amino acid oxidase. *Biochemistry* **1997**, *36*, 5624–5632. [CrossRef] [PubMed]

52. Sasabe, J.; Chiba, T.; Yamada, M.; Okamoto, K.; Nishimoto, I.; Matsuoka, M.; Aiso, S. D-Serine is a key determinant of glutamate toxicity in amyotrophic lateral sclerosis. *EMBO J.* **2007**, *26*, 4149–4159. [CrossRef] [PubMed]

53. Miladinovic, T.; Nashed, M.G.; Singh, G. Overview of Glutamatergic Dysregulation in Central Pathologies. *Biomolecules* **2015**, *5*, 3112–3141. [CrossRef] [PubMed]

54. Yang, S.; Qiao, H.; Wen, L.; Zhou, W.; Zhang, Y. D-serine enhances impaired long-term potentiation in CA1 subfield of hippocampal slices from aged senescence-accelerated mouse prone/8. *Neurosci. Lett.* **2005**, *379*, 7–12. [CrossRef] [PubMed]

behavioral sciences

MDPI

Article

Relation of Structural and Functional Changes in Auditory and Visual Pathways after Temporal Lobe Epilepsy Surgery

Margarita Minou Báez-Martín [1,*], Lilia Maria Morales-Chacón [1], Iván García-Maeso [1],
Bárbara Estupiñán-Díaz [1], María Eugenia García-Navarro [1], Yamila Pérez Téllez [1],
Lourdes Lorigados-Pedre [1], Nelson Quintanal-Cordero [1], Ricardo Valdés-Llerena [1],
Judith González González [1], Randis Garbey-Fernández [1], Ivette Cabrera-Abreu [1],
Celia Alarcón-Calaña [1], Juan E. Bender del Busto [1], Rafael Rodríguez Rojas [2],
Karla Batista García-Ramó [2] and Reinaldo Galvizu Sánchez [3]

[1] Epilepsy Surgery Program, International Center for Neurological Restoration, 25th Ave, No 15805,
PC 11300 Havana, Cuba; lily@neuro.ciren.cu (L.M.M.-C.); ivangarciamaeso@gmail.com (I.G.-M.);
baby@neuro.ciren.cu (B.E.-D.); marugeniagarcian@gmail.com (M.E.G.-N.); yamilaperez86@yahoo.es (Y.P.T.);
lourdesl@neuro.ciren.cu (L.L.-P.); nquintanal@neuro.ciren.cu (N.Q.-C.); rvaldes@neuro.ciren.cu (R.V.-L.);
judith@neuro.ciren.cu (J.G.G.); randis0770@gmail.com (R.G.-F.); ivettec@neuro.ciren.cu (I.C.-A.);
calarcon@neuro.ciren.cu (C.A.-C.); jebender@infomed.sld.cu (J.E.B.d.B.)
[2] Imaging Department, International Center for Neurological Restoration, 25th Ave, No 15805,
PC 11300 Havana, Cuba; rrguezrojas@gmail.com (R.R.R.); kbatista@neuro.ciren.cu (K.B.G.-R.)
[3] Clinical Department, International Center for Neurological Restoration, 25th Ave, No 15805,
PC 11300 Havana, Cuba; rgalvizu@infomed.sld.cu
* Correspondence: minou@neuro.ciren.cu; Tel.: +53-7273-6923

Received: 14 August 2018; Accepted: 5 October 2018; Published: 12 October 2018

Abstract: Auditory and visual pathways may be affected as a consequence of temporal lobe epilepsy surgery because of their anatomical relationships with this structure. The purpose of this paper is to correlate the results of the auditory and visual evoked responses with the parameters of tractography of the visual pathway, and with the state of connectivity between respective thalamic nuclei and primary cortices in both systems after the surgical resection of the epileptogenic zone in drug-resistant epileptic patients. Tractography of visual pathway and anatomical connectivity of auditory and visual thalamus-cortical radiations were evaluated in a sample of eight patients. In general, there was a positive relationship of middle latency response (MLR) latency and length of resection, while a negative correlation was found between MLR latency and the anatomical connection strength and anatomical connection probability of the auditory radiations. In the visual pathway, significant differences between sides were found with respect to the number and length of tracts, which was lower in the operated one. Anatomical connectivity variables and perimetry (visual field defect index) were particularly correlated with the latency of P100 wave which was obtained by quadrant stimulation. These results demonstrate an indirect functional modification of the auditory pathway and a direct traumatic lesion of the visual pathway after anterior temporal lobectomy in patients with drug resistant epilepsy.

Keywords: auditory evoked responses; connectivity; drug-resistant epilepsy; temporal lobectomy; tractography; visual evoked potentials

1. Introduction

The natural history of mesial temporal lobe epilepsy (mTLE) is variable; among 20% and 40% of patients have pharmacologically intractable seizures despite more than ten new antiepileptic drugs being added to the market in recent years [1–4]. This refractoriness to medication leads to a progressive deterioration from the neurobiological, psychological and social point of view of the patients. Hence, surgical resection of the epileptogenic zone continues to be an important and effective therapeutic option for people with mTLE, with seizures totally disappearing in around 70–85% of cases [3,5,6]. However, due to the morbidity of the procedure, the risk-benefit balance, as well as secondary effects, should be considered. Particularly, the surgery of the temporal lobe may involve structures related to sensory processing. From the anatomical and functional point of view, the temporal lobe is closely related to the auditory and visual pathways before its projection to the temporal operculum and the calcarine sulcus, respectively [7]. Thus, these pathways can be directly or indirectly affected as a consequence of surgical treatment.

Visual field defects may appear as a consequence of direct injury of the optic radiations during the surgical procedure (superior quadrantanopia contralateral to the side of resection), particularly the Meyer's loop which runs in the vicinity of the temporo-mesial structures [8–13]. However, less is known about possible modifications that may occur in the auditory pathway [6,14].

In previous papers, we studied the variations of auditory and visual evoked potentials in patients with temporal lobe epilepsy that were submitted to surgical resection of the epileptogenic area [15,16]. Auditory brainstem response, auditory middle latency response and partial field visual evoked potentials were evaluated in particular. Additionally, some anatomical characteristics of the resected tissue and its relationship with the auditory and visual pathways were evaluated in a sample of these patients. A high coincidence rate between quadrant visual evoked potentials and perimetry was found, particularly in the superior quadrant that was contralateral to surgery. The volumetric study also showed that the injury of structures were probably related to the visual field defect. However, the results of the auditory evoked potentials and their relation to the volumetric findings pointed towards indirect damage of the auditory pathway, secondary to surgery. The aim of this paper is to correlate the results of the quadrant visual evoked potentials with the parameters of tractography of the visual pathway, and the results of the auditory and visual evoked responses with the state of connectivity between respective thalamic nuclei and primary cortices in both systems after the surgical resection of the epileptogenic zone.

2. Materials and Methods

2.1. Subjects

The study sample consisted of twenty-seven patients who were evaluated at the Telemetric Unit of the International Center for Neurological Restoration in Havana, Cuba in a prospective study. These patients were referred to this Unit from specialized consultations throughout the country once their intractability to medications was defined. They had crisis for at least two years with the use of at least two major antiepileptic drugs (Carbamazepine, Diphenylhydantoin, Valproate, Phenobarbital, Primidone) at the maximum tolerated doses for adequate periods of time. The inclusion criteria were:

- Patients with mTLE evaluated according to the institution's protocols and who were candidates for resective surgery according to agreed criteria.
- Patients who met the drug-resistance criteria.
- Patients who gave their informed consent to participate in the investigation.

Presurgical evaluation was performed, including the localization of the ictal zone by combining Electroencephalogram-video, ictal and interictal SPECT (Single Photon Emission Computed Tomography), qualitative and quantitative MRI and neuropsychological tests. Clinical and demographical data of the patients and controls are shown in Table 1.

The most frequently used drugs for treatment before and after surgical procedure were carbamazepine, clonazepam and valproate, and this did not change for at least two years after the surgical procedure. By the time the study began, 88.8% of patients had taken more than one medication.

Table 1. Clinical and demographical characteristics of the sample.

	N	Sex		Age Years μ (SD)	Duration of Illness Years μ (SD)	Temporo-Mesial Sclerosis N	Focal Cortical Dysplasia		
		M	F				IIIa	IIIb	No
Patients									
Left lobectomy	14	5	9	33.07 (9.06)	21.14 (9.08)	14	8	2	4
Right lobectomy	13	7	6	34.69 (5.28)	22.53 (12.14)	13	9	1	3
Healthy subjects	16	8	8	33.68 (7.96)	-	-	-		

μ: mean value; SD: standard deviation.

All patients underwent standard anterior temporal lobectomy guided by electrocorticography. Tissue samples that were obtained during surgery were histopathologically studied. Also, focal cortical dysplasias were classified according to Blumcke's criteria [17] (Table 1).

Refraction defects were tested and corrected in every subject before the evaluation of the visual system. Additionally, hearing threshold level was evaluated.

All subjects were right-handed and gave their consent to participate in the study.

2.2. Electrophysiological Tests

Auditory brainstem response (ABR), middle latency response (MLR) and partial field visual evoked potentials were measured in all of the patients and healthy subjects.

The recording conditions of the auditory and visual electrophysiological tests are summarized in Table 2.

The records were obtained with a Neuropack four-mini and a Neuropack M1 device (Nihon Khoden, Tokyo, Japan).

Table 2. Recording conditions of auditory responses and visual evoked potentials.

Parameters	ABR	MLR	VEP
Analysis time (ms)	10	100	300
Filters (Hz)	100–3000	20–1000	1–100
Stimulus frequency (Hz)	10	5	1
Maximal intensity (dBnHL)	105	90	-
Average responses	2000	500	30
Sensibility (μV/div)	5	20	20
Recording electrodes	A1, A2	Cz	O1, Oz, O2
Reference electrode	Cz	A1-A2	Fz
Ground electrode	Fpz	Fpz	A1
Stimulation mode	Monoaural	Binaural	Quadrants

ABR: auditory brainstem response; MLR: middle latency response; VEP: visual evoked potentials.

The auditory stimuli were 0.1 ms alternating clicks that were delivered through a headphone (DR-531B-7, Elegas Acous Co. Ltd., Tokyo, Japan).

The visual stimuli consisted of a pattern-reversal checkerboard (black and white checks) with a fixation center-point, 16' size and high contrast. They were presented on a VD-401A monitor.

2.3. Resection Measure

The absolute length of the lateral and mesial resected tissue was measured by the neurosurgeon using slices of MRI in T1, T2 and FLAIR (Fluid-Attenuated Inversion Recovery) six months after surgery. The lateral (neocortical) aspect included the distance (in mm) between the posterior edge of the internal table and the anterior limit of the resected area, considering the middle temporal gyrus in the axial slices of the MRI, whereas the mesial aspect was calculated from axial slices in parallel with the preserved hippocampus. Two tangential lines were drawn: one between the posterior border of the resection and the contralateral hippocampus, and the other between the anterior limit of the preserved hippocampus and the side of resection. The distance between these tangential lines was the mesial longitude.

2.4. Perimetry

The requirements for the perimetric evaluation can be reviewed in Báez Martín et al. 2010 (16). Moreover, the area of the superior quadrant that was ipsilateral to the resection was used as a reference of the subject itself [18], and the calculation of the visual field defect index (VFD-I) was carried out using the following expression:

VFD-I = 1 − (preserved area in the contralateral superior quadrants/preserved area in the ipsilateral superior quadrants), so that a high index corresponded to a greater visual defect.

2.5. Diffusion Tensor Imaging (DTI)

The diffusion neuroimaging studies were performed in eight patients, six with right lobectomy and two with left. By using the standard gradient direction scheme (twelve weighted diffusion images and one b = 0 image), the diffusion images were acquired by an echo-planar sequence (EPI) with the following parameters: b = 1200 s/mm^2; FOV = 256 × 256 mm^2; acquisition matrix = 128 × 128 mm^2; corresponding to a spatial resolution 2 × 2 mm^2; TE/TR = 160 ms/7000 ms.

For the processing of the images, the diffusion spectrum imaging (DSI) Studio software (dsi-studio.labsolver.org/dsi-studio-download) was used and the DTI method that was proposed by Basser et al. which characterizes the main direction of fiber diffusion in the brain [19].

2.6. Tractography

Quantitative measurements, such as fractional anisotropy (FA), apparent diffusion coefficient (ADC) as well as axial and radial diffusivity values were obtained. These measures were analyzed through predetermined areas, according to regions of interest (ROIs) that were positioned following anatomical guides. In the case of optic radiation, the ROI was drawn on the FA map of each subject individually, also evaluating the occipital cortex (area 17 of Brodmann) as an independent ROI. This analysis was not possible for the auditory pathway due to the poor visualization of the thalamus-cortical segment, which did not allow drawing the ROI on the FA map.

2.7. Brain Anatomical Connectivity

The condition of the connections between the thalamus and the cortex for the visual and auditory pathways was evaluated. In the case of the visual pathway, anatomical connectivity was measured between the lateral geniculate body and the primary visual cortex (area 17 of Brodmann). For the auditory pathway, the same analysis was performed between the medial geniculate body and the primary auditory cortex (area 41 of Brodmann). A methodology that was based on graph theory that was proposed by Iturria et al. (2007) was implemented to describe the specific connections between the different regions of gray matter. The method consists of three essential steps: (1) definition of a Brain Graph model in which each voxel is considered as a node of a non-directed weighted graph; (2) use of an iterative algorithm to find the route of maximum probability between two nodes and the subsequent definition of the anatomical connectivity measure between them; (3) definition of three anatomical connectivity measures between different gray matter regions: anatomical connection strength (ACS), anatomical connection probability (ACP) and anatomical connection density (ACD) [20].

2.8. Statistics

Data analysis included the comparison between hemispheres for tractography and connectivity six months after surgery (*t* test for dependent samples). Normalization of the electrophysiological data was carried out using the mean and standard deviation values of the control group for correlation studies. The relationship among electrophysiological tests, perimetry, length of resection, tractography and anatomical connectivity was measured by the Pearson correlation test. Differences were considered significant if $p < 0.05$ for all tests (Statistica 8.0.360 Copyright StatSoft. Inc., Tulsa, OK, USA, 1984–2011).

2.9. Ethical Considerations

All the procedures followed the rules of the Declaration of Helsinki of 1975 for human research, and the study was approved by the scientific and ethics committee from the International Center for Neurological Restoration (CIREN37/2012).

3. Results

3.1. Auditory Pathway

As already noted, it was not possible to evaluate the tracts corresponding to the auditory radiations, however it was possible to study their connections. The results (mean and standard deviation) of the variables that were considered for this study are summarized in Table 3.

Table 3. Auditory radiations variables of DTI (Diffusion Tensor Imaging).

Variables	Ipsilateral Auditory Radiation		Contralateral Auditory Radiation		
	μ	SD	μ	SD	*p*
Density of connection	0.193	0.0773	0.251	0.0660	0.026 *
Density of connection (FA)	0.068	0.0293	0.093	0.0189	0.0800
Density of connection (MD)	0.080	0.0306	0.112	0.0261	0.0652
Anatomical Connection probability	0.422	0.1613	0.498	0.1172	0.3981
Anatomical Connection probability (FA)	0.181	0.0992	0.210	0.0372	0.4783
Anatomical Connection probability (MD)	0.193	0.0965	0.230	0.0565	0.4317
Anatomical connection strength	726.900	402.7487	1214.176	330.9955	0.0346 *
Anatomical connection strength (FA)	257.677	149.3024	449.342	100.6329	0.020 *
Anatomical connection strength (MD)	302.172	161.8940	542.212	135.5526	0.0214 *

FA: fractional anisotropy MD: media diffusivity. μ: media; SD: standard deviation. * Statistically significant differences between sides (*t* test for dependent samples, $p < 0.05$).

Structure-Function Relationships

For the correlation analysis with the anatomical variables, using the normalized data, both groups of patients were united, considering whether the pathway to be evaluated was ipsi or contralateral to the resection.

ABR: There were no statistically significant correlation between the variables of the ABR and the length of resection (Pearson correlation test, $p > 0.05$). There was a negative correlation between the connection density and the latency of the wave III that was ipsilateral to the resection (Pearson correlation test, $p < 0.05$, r = −0.76) only in the analysis of the connectivity in the auditory radiations. This showed a significant prolongation of latency in the group with right lobectomy.

MLR: As a whole, we found a positive relationship between the latency of the MLR components and the length of the resection; whereas a negative one was observed with the connectivity of the auditory thalamus-cortical radiation.

Table 4 summarizes the results of the correlation analysis between the electrophysiological variables of the MLR and the anatomical studies (Pearson correlation test, $p < 0.05$).

Table 4. Correlations between middle latency response and anatomical variables.

	Na Latency	Pa Latency	Nb Latency	Na Amplitude
Mesial length of resection	+0.13	+0.51 *	+0.48 *	−0.39
Neocortical length of resection **	+0.61 *	+0.31	+0.21	+0.16
Anatomical connection probability (FA) **	−0.88 *	−0.86	−0.89 *	+0.92 *
Anatomical connection strength (FA) *	−0.87	−0.89 *	−0.89 *	0.83

Note: The table refers to the correlation of the anatomical variables that were ipsilateral to the resection with the normalized values of middle latency response recorded in Cz. The + and − signs indicate whether the relationship is positive or negative, and the numbers correspond to the r of each correlation (Pearson correlation test, $p < 0.05$). FA: fractional anisotropy. ** Patients with right temporal lobectomy. * Statistically significant values.

The latency of Na correlated positively with the length of neocortical resection, however only in the group of patients with right lobectomy (Pearson correlation test, $p < 0.05$, r = 0.61). This group is precisely the one that showed a prolongation of latency of this component after resection, which reached statistical significance twelve months after surgery (*t* test, $p < 0.05$). It was also the group that showed the most extensive resection in the temporal neocortex (average 42.81 mm). The latency of the components Pa and Nb correlated positively with the mesial length of the resection in the whole group of operated patients (Pearson correlation test, $p < 0.05$, r = 0.51 and r = 0.48, respectively).

In a previous report, it described the positive correlation between the amplitude of Na and Pa waves and the resected volume in the middle temporal pole (Pearson correlation test $p < 0.05$, r = 0.84 and r = 0.65, respectively), especially in left temporal lobectomy patients. On the contrary, negative correlations were previously observed between the amplitude of these components and the residual indexes of the middle temporal pole (Pearson correlation test, $p < 0.05$, Na: r = −0.89 y Pa: r = −0.79), inferior temporal gyrus and amygdala (Pearson correlation test, $p < 0.05$, only Pa: r = −0.67 y r = −0.59, respectively) [15].

3.2. Visual Pathway

The results (mean and standard deviation) of the variables that were considered for the white matter tracts of optic radiation in both hemispheres are summarized in Table 5.

Table 5. DTI values in optic radiation and occipital cortex. Tractography.

		Ipsilateral		Contralateral		
		μ	SD	μ	SD	*p*
OR	Number of tracts **	356.625	185.997	578.500	155.902	0.0041 *
	Length of tracts	19.671	7.139	26.870	7.557	0.0307 *
	Volume of tracts	3167.500	1339.976	4368.750	1462.917	0.1988
	FA	0.403	0.127	0.499	0.031	0.0587
	ADC	0.001	0.000	0.001	0.000	0.1448
	AD	0.001	0.000	0.001	0.000	0.3502
	RD	0.001	0.000	0.001	0.000	0.1228
OC	Number of tracts	1961.87	832.04	2971.25	661.71	0.0017 *
	Length of tracts	14.986	4.748	18.190	2.669	0.1127
	Volume of tracts	7396.25	3425.97	9145.00	2655.12	0.0499 *
	FA	0.348	0.083	0.388	0.099	0.0026 *
	ADC	0.001	0.000	0.001	0.000	0.1062
	AD	0.001	0.000	0.001	0.000	0.7131
	RD	0.001	0.000	0.001	0.000	0.0741

OR: optic radiation; OC: occipital cortex. FA: fractional anisotropy; ADC: apparent diffusion coefficient; AD: axial diffusivity; RD: radial diffusivity. μ: average; SD: standard deviation. ** Right lobectomy patients. (*t* test for dependent samples, $p < 0.05$). * Statistically significant differences between sides. When comparing the optic radiations of both sides, statistically significant differences were evidenced for the whole group of patients with respect to the length of the tracts (*t* test for dependent samples, $p = 0.0307$) being smaller on the operated side. When we limited the analysis to the group with right lobectomy, the number of tracts was also significantly different (*t* test for dependent samples, $p = 0.0041$).

The comparative analysis of the tracts that were related to the occipital cortex showed differences between the sides that were statistically significant for the number of tracts, volume of tracts and fractional anisotropy (*t* test for dependent samples), which was also lower on the operated side.

Table 6 shows the results (mean and standard deviation) of the variables that were considered in the study of anatomical connectivity. The comparisons between the sides showed lower values of density, probability and strength of the anatomical connection on the side of the resection. However, these differences did not reach statistical significance (*t* test for dependent samples, $p > 0.05$).

Table 6. DTI values in optic radiation. Anatomical connectivity.

	Ipsilateral		Contralateral		
	μ	SD	μ	SD	*p*
Anatomical connection density	0.155	0.0733	0.174	0.0757	0.6104
Anatomical connection density (FA)	0.050	0.0271	0.069	0.0323	0.2115
Anatomical connection density (MD)	0.059	0.0306	0.075	0.0305	0.3426
Anatomical connection probability	0.387	0.1810	0.404	0.1292	0.8029
Anatomical connection probability (FA)	0.134	0.0711	0.174	0.0512	0.2581
Anatomical connection probability (MD)	0.178	0.0885	0.181	0.0537	0.9418
Anatomical connection strength	890.987	450.569	1239.440	610.7670	0.2174
Anatomical connection strength (FA)	290.970	171.595	487.988	260.4703	0.0958
Anatomical connection strength (MD)	342.933	188.255	531.343	253.9019	0.1243

FA: fractional anisotropy; MD: medium diffusivity. μ: average; SD: standard deviation (*t* test for dependent samples, $p > 0.05$).

Structure-Function Relationships

For the analysis of correlation with the anatomical studies, both groups of patients were united considering whether the quadrant to be evaluated was ipsi or contralateral to the resection, and standardized data of the quadrant-visual evoked potentials was used. The longest latency (maximum latency) and the smallest amplitude (minimum amplitude) of the P100 wave were taken from the pair of affected quadrants.

Table 7 summarizes the results of the correlation analysis that had statistical significance among the electrophysiological variables, the perimetry and the anatomical studies (Pearson correlation test, $p < 0.05$).

Table 7. Correlations with statistical significance between the anatomical variables, quadrant visual evoked potentials and the perimetry.

(A) Optic Radiations	P100 Latency			P100 Amplitude			Perimetry (VFD-I)
Electrodes	O1	Oz	O2	O1	Oz	O2	
Perimetry	+0.53	+0.50	+0.50				
Neocortical length of resection	+0.61	+0.60	+0.59				+0.72
FA *	−0.96	−0.98	−0.97				−0.73
ADC *	+0.98	+0.97	+0.98				
Axial diffusivity *	+0.99	+0.99	+0.99				
Radial diffusivity *	+0.99	+0.99	+0.99				
Connection density	−0.79	−0.78	−0.78		+0.76		
Connection probability	−0.89	−0.87	−0.86				
Connection strength	−0.88	−0.88	−0.89		+0.74		

Table 7. *Cont.*

(B) Occipital Cortex	P100 Latency			P100 Amplitude			Perimetry (VFD-I)
Electrodes	O1	Oz	O2	O1	Oz	O2	
Number of tracts						+0.86	
Volume of tracts				+0.87	+0.94	+0.80	
FA					+0.82		−0.71

Note: The table refers to the correlation of the anatomical variables with the normalized values of the visual evoked potential by quadrants (maximum latency and minimum amplitude) in the contralateral superior quadrant to the resection. The + and − signs indicate whether the relationship is positive or negative, and the numbers correspond to the r of each correlation (Pearson correlation test, $p < 0.05$). * Patients with right lobectomy, right eye. VFD-I: visual field defect index. FA: fractional anisotropy; ADC: apparent diffusion coefficient.

Figure 1 shows an example of the tractography of optic radiation in a patient with right lobectomy, where the reduction in the number of fibers on that side is appreciated and whose perimetry showed a superior left quadrantanopia.

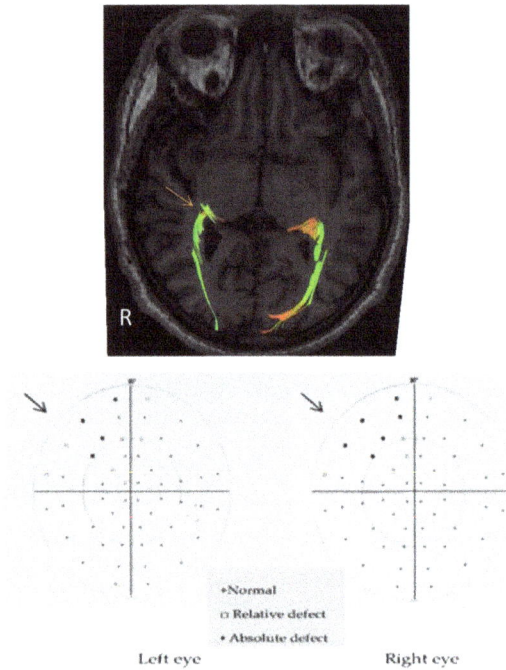

Figure 1. Tractography of optic radiations and perimetry of a patient with right lobectomy. Arrows indicate the pathological optic radiation (tractography) and the affected quadrants (perimetry). R: right side of the image.

4. Discussion

Epilepsy surgery as a therapeutic option for patients with drug-resistant mTLE produced structural and functional changes in the auditory and visual sensory systems, which were detected by the use of electrophysiological techniques and in correspondence with the structural variations that were evidenced with neuroimaging techniques.

The relationships that were found between the brainstem exploration (ABR) on the same side of the resection and the density of the connection in the auditory pathway could correspond to the known fact that the fibers which participate in the efferent control that are exerted by the cortex on the lower levels of the auditory pathway (mesencephalon-pons-cochlear receptor) are mostly ipsilateral [21].

It is known that the primary auditory cortex, preserved during this surgery, inhibits the conduction of the ascending impulses by means of the sensory control system, modifying the activity in the relay stations of the pathway (inferior colliculi and olive-cochlear complex) ipsilaterally, while the stimulation of secondary auditory areas enhances the ascending impulses [21]. These latter areas could be influenced by the removed zone, especially in patients with right lobectomy where the resection is wider. As a result of this excitatory-inhibitory imbalance, the inhibition of the aference on the same side of the surgery dominates, which is expressed by a delay in conduction at the subcortical levels of the pathway.

The magnitude of the resection of mesial structures, which is greater in the group with left temporal lobectomy, had greater repercussions on the thalamus-cortical components of the MLR (Pa, Nb), while the extension of the resection in its neocortical lateral aspect exerted a greater influence on the most caudal component (Na), especially in the group with right lobectomy. All these results speak in favor of a possible functional relationship between the resected structures and the generators of these components, with a variable behavior depending on whether the operated hemisphere is right or left.

The study of the anatomical connectivity showed an inverse relationship of the probability and strength of connection (particularly its FA) with the latency of Pa and Nb for the auditory radiation of the operated hemisphere. Interestingly, the Na component was related to the connection probability: the higher the latency and the lower the amplitude, the lower the connection probability (see Table 4).

These results suggest the existence of a modification of the functional state in the auditory pathway, which is given by a lower speed of propagation of the nerve impulse in correspondence with a reduction in the flow of information from the medial geniculate body to the auditory cortex and a lower functional relationship among these structures, with repercussion to other lower levels in the neuroaxis (inferior colliculi-superior olive complex).

However, if we take into account that during the postsurgical evaluation, the amplitude analysis of the components of the MLR that were far from decreasing showed a tendency to increase at 12 and 24 months after the resection [15], we could be in the presence of a "paradoxical" phenomenon of release, characterized by a long-term functional improvement once the epileptogenic zone is removed.

The possibility that this change is dependent on a physical factor, such as the reduction of impedance at the surgical site, does not seem feasible, given that the increase in amplitude was progressive over time.

The results of the auditory examination confirm what has been proposed in previous studies about the occurrence of long-term neuroplastic changes secondary to an indirect effect of the resection of temporal lobe structures [15]. Although our data seem to point to a possible effect of the surgery in the auditory pathway, these measures are not fully independent. In summary, there are functional implications of mTLE surgery on auditory processing. Concerning the visual pathway, the damage of Meyer's loop after temporal lobectomy [9,10,22–30] is well known, with the consequent visual field defect (homonymous superior quadrantanopia contralateral to the resected side) that can limit activities of daily life like driving [28,31].

Unlike the postulates that were used in the evaluation of the auditory pathway, in the case of the visual pathway, the sequelae of the anterior temporal lobectomy do constitute, a consequence of the direct aggression of the optical radiation fibers, dependent on the surgical procedure employed [32] and the distance between the tip of the Meyer loop and the posterior edge of the temporal pole [33]. Its anatomical disposition predisposes for this type of affectation to occur, especially in patients with right lobectomy where there is no risk of damaging eloquent areas for language, with which more generous removals from the epileptogenic zone can be carried out guided by trans-operative electrocorticography.

However, none of these studies have considered the relationship between functional variations and anatomical changes of the visual pathway.

The latency of the P100 wave that was obtained with quadrant stimulation correlated with the fiber damage in optic radiation, which was demonstrated by the effect of the extension of the resection in the neocortex of the temporal lobe (positive) and the value of the connectivity between the lateral geniculate body and the occipital cortex (negative) on this electrophysiological variable. This relationship was also evident between the latency of the P100 wave and the results of the perimetry (positive) for the three recording electrodes.

The amplitude of P100 correlated with the volume of the tracts that were related to the occipital cortex. The density and strength of the connection between the regions of interest had a positive relationship with this variable, selectively for the O2 electrode (right occipital cortex), which seems to be congruent given that most of the patients that were included in the anatomical study of the tracts had resection of the anterior temporal lobe of the right side (6 of 8). Furthermore, subjects with the higher volume of resected tissue showed the most delayed latencies and the lowest P100 amplitude [16].

Note that perimetry, classically considered the "gold standard" for the evaluation of visual field defects, also had a statistically significant relationship with the neocortical length of the resection and with the values of fractional anisotropy, both from the optic radiation and the occipital cortex (See Table 7).

When evaluating the tractography variables in optic radiation, we found no statistically significant relationship with the remaining measurements in the studied subjects (Pearson correlation test, $p > 0.05$).

However, when considering the side of the resection and the responses of each eye separately, a negative correlation of the P100 latency in the contralateral superior quadrant of the ipsilateral eye was found with the values of fractional anisotropy, and a positive correlation with the ADC and diffusivity (axial and radial) in patients with right lobectomy. This result suggests, as it did with visual potential and perimetry, that there is a higher probability of damaging fibers from the eye on the same side of the resection that derive from the temporal retina (nasal visual field) and that run more ventrally in the Meyer's loop, causing the incongruous visual field defects [8].

Finally, it is worth noting that we found a statistically significant negative correlation of the tract volume and the fractional anisotropy of the occipital cortex of the operated side with the P100 latency in the contralateral superior quadrant (Pearson correlation test, $p < 0.05$, r = -0.87 and r = -0.89 respectively), and with the amplitude of P100 in the same quadrant (Pearson's correlation test, $p < 0.05$, r = 0.94 and r = 0.92, respectively). The above was valid only when the analysis of the patients with right temporal lobectomy was performed. Once again, the results point to a greater anatomical and functional compromise of optic radiation when it comes to the non-dominant hemisphere for language.

In summary, the highest values of P100 latency in the contralateral superior quadrant indicate probable myelin damage in the optic radiation, which is supported by neuroimaging studies of diffusion (tractography and connectivity) and in correspondence with the perimetric results [16]. All of these findings confirm the existence of a partial and selective dysfunction of the visual pathway, in particular of the optic radiations on the side of the temporal lobectomy, which is expressed as a superior homonymous quadrantanopia contralateral to the resection, which is imperceptible for most patients with epilepsy that are resistant to medication and subjected to this surgical procedure.

These visual sequels could be minimized with transoperative monitoring guided by tractography of optic radiations, a method that is not available today in all groups that perform this type of surgery.

Among the limitations of this study is the absence of presurgical references of the magnetic resonance images in both pathways, as well as the difficulties to evaluate the images of the auditory radiations.

5. Conclusions

Taken as a whole, these results demonstrate an indirect functional modification of the auditory pathway and a direct traumatic lesion of the visual pathway after anterior temporal lobectomy in patients with pharmaco-resistant epilepsy.

The novelty of our study lies in the combined use of electrophysiological and imaging techniques to evaluate sensory pathways that are related to resected tissue in patients undergoing this surgical procedure. Undoubtedly, the results of tractography and structural connectivity are in line with the findings of the functional tests, which further strengthens our study.

Author Contributions: M.M.B.-M. and L.M.M.-C. conceived and designed the experiments and wrote the paper; I.G.-M. and N.Q.-C. made the surgical procedures; R.V.-L. and R.G.-F. made the anesthetics procedures; Y.P.T., I.C.-A. and C.A.-C. performed the electrophysiological tests; J.G.G., J.E.B.d.B., and R.G.S. made the clinical evaluation; M.E.G.-N. performed the neuropsychological evaluation; K.B.G.-R. and R.R.R. made the image processing; B.E.-D. and L.L.-P. performed the anatomopathological procedures.

Funding: This research received no external funding.

Acknowledgments: We thank Odalys Morales-Chacon for the English revision of the manuscript and Jorge Bergado-Rosado for his contributions.

Conflicts of Interest: The authors declare no conflict of interest.

References

1. Gronich, G.; Arno, L.; Dualib, K. Neurofisiologia nao-invasiva (EEG) das diferentes síndromes epilépticas. In *Tratamento clínico e cirúrgico das epilepsias de difícil controle*; Cukiert, A., Ed.; Lemos: Sao Paulo, Brazil, 2002; p. 157.
2. Hauser, W. The natural history of temporal lobe epilepsy. In *Epilepsy Surgery*; Luders, H., Ed.; Raven Press: New York, NY, USA, 2000; pp. 133–141.
3. Viera, O., Jr. Tratamento cirúrgico da epilepsia do lobo temporal. In *Tratamento Clínico e Cirúrgico das Epilepsias de Difícil Controle*; Cukiert, A., Ed.; Lemos: Sao Paulo, Brazil, 2002; pp. 269–289.
4. Bonilha, L.; Marzt, G.U.; Glazier, S.S.; Edwards, J.C. Subtypes of medial temporal lobe epilepsy: Influence on temporal lobectomy outcomes? *Epilepsia* **2012**, *53*, 1–6. [CrossRef] [PubMed]
5. Bazil, C.; Morrell, M.J.; Peddley, T. Epilepsy. In *Merritt's Neurology*; Rowland, L., Ed.; Lippincott Williams & Wilkins: Philadelphia, PA, YSA, 2005; pp. 990–1008.
6. Karceski, S.; Morrell, M. Principles of epilepsy management: Diagnosis and treatment. In *Neurological Therapeutics: Principles and Practice*; Noseworthy, J.H., Ed.; Informa Healthcare: Rochester, NY, USA, 2006; pp. 341–354.
7. Sindou, M.; Guenot, M. Surgical anatomy of the temporal lobe for epilepsy surgery. *Adv. Tech. Stand. Neurosurg.* **2003**, *28*, 315–343. [PubMed]
8. Hughes, T.S.; Abou-Khalil, B.; Lavin, P.J.; Fakhoury, T.; Blumenkopf, B.; Donahue, S.P. Visual field defects after temporal lobe resection: A prospective quantitative analysis. *Neurology* **1999**, *53*, 167–172. [CrossRef] [PubMed]
9. James, J.S.; Radhakrishnan, A.; Thomas, B.; Madhusoodanan, M.; Kesavadas, C.; Abraham, M.; Menon, R.; Rathore, C.; Vilanilam, G. Diffusion tensor imaging tractography of Meyer's loop in planning resective surgery for drug-resistant temporal lobe epilepsy. *Epilepsy Res.* **2015**, *110*, 95–104. [CrossRef] [PubMed]
10. Jeelani, N.U.; Jindahra, P.; Tamber, M.S.; Poon, T.L.; Kabasele, P.; James-Galton, M.; Stevens, J.; Duncan, J.; McEvoy, A.W.; Harkness, W.; et al. 'Hemispherical asymmetry in the Meyer's Loop': A prospective study of visual-field deficits in 105 cases undergoing anterior temporal lobe resection for epilepsy. *J. Neurol. Neurosurg. Psychiatry* **2010**, *81*, 985–991. [CrossRef] [PubMed]
11. Nilsson, D.; Malmgren, K.; Rydenhag, B.; Frisen, L. Visual field defects after temporal lobectomy—Comparing methods and analysing resection size. *Acta Neurol. Scand.* **2004**, *110*, 301–307. [CrossRef] [PubMed]
12. Pathak-Ray, V.; Ray, A.; Walters, R.; Hatfield, R. Detection of visual field defects in patients after anterior temporal lobectomy for mesial temporal sclerosis-establishing eligibility to drive. *Eye (Lond.)* **2002**, *16*, 744–748. [CrossRef] [PubMed]

13. Mandelstam, S.A. Challenges of the Anatomy and Diffusion Tensor Tractography of the Meyer Loop. *AJNR Am. J. Neuroradiol.* **2012**, *33*, 1204–1210. [CrossRef] [PubMed]
14. Japaridze, G.; Kvernadze, D.; Geladze, T.; Kevanishvili, Z. Auditory brain-stem response, middle-latency response, and slow cortical potential in patients with partial epilepsy. *Seizure* **1997**, *6*, 449–456. [CrossRef]
15. Baez-Martin, M.M.; Morales Chacon, L.M.; Garcia-Maeso, I.; Estupiñan-Diaz, B.; Lorigados-Pedre, L.; Garcia, M.E.; Galvizu, R.; Bender, J.E.; Cabrera-Abreu, I.; Perez-Tellez, Y.; et al. Temporal lobe epilepsy surgery modulates the activity of auditory pathway. *Epilepsy Res.* **2014**, *108*, 748–754. [CrossRef] [PubMed]
16. Baez Martin, M.M.; del Carmen Perez, T.Y.; Morales Chacon, L.M.; Diaz, B.E.; Trapaga-Quincoses, O.; Maeso, I.G.; Bender, J.E.; Galvizu, R.; Garcia, M.E.; Abreu, I.C.; et al. Innovative evaluation of visual field defects in epileptic patients after standard anterior temporal lobectomy, using partial field visual evoked potentials. *Epilepsy Res.* **2010**, *90*, 68–74. [CrossRef] [PubMed]
17. Blumcke, I.; Aronica, E.; Urbach, H.; Alexopoulos, A.; Gonzalez-Martinez, J.A. A neuropathology-based approach to epilepsy surgery in brain tumors and proposal for a new terminology use for long-term epilepsy-associated brain tumors. *Acta Neuropathol.* **2014**, *128*, 39–54. [CrossRef] [PubMed]
18. Daga, P.; Winston, G.; Modat, M.; White, M.; Mancini, L.; Cardoso, M.J.; Symms, M.; Stretton, J.; McEvoy, A.W.; Thornton, J.; et al. Accurate localization of optic radiation during neurosurgery in an interventional MRI suite. *IEEE Trans. Med. Imaging* **2012**, *31*, 882–891. [CrossRef] [PubMed]
19. Basser, P.J.; Mattiello, J.; LeBihan, D. Estimation of the effective self-diffusion tensor from the NMR spin echo. *J. Magn. Reson. B* **1994**, *103*, 247–254. [CrossRef] [PubMed]
20. Iturria-Medina, Y.; Canales-Rodriguez, E.J.; Melie-Garcia, L.; Valdés-Hernández, P.A.; Martínez-Montes, E.; Alemán-Gómez, Y.; Sánchez-Bornot, J.M. Characterizing brain anatomical connections using diffusion weighted MRI and graph theory. *Neuroimage* **2007**, *36*, 645–660. [CrossRef] [PubMed]
21. Khalfa, S.; Bougeard, R.; Morand, N.; Veuillet, E.; Isnard, J.; Guenot, M.; Ryvlin, P.; Fischer, C.; Collet, L. Evidence of peripheral auditory activity modulation by the auditory cortex in humans. *Neuroscience* **2001**, *104*, 347–358. [CrossRef]
22. Chen, X.; Weigel, D.; Ganslandt, O.; Buchfelder, M.; Nimsky, C. Prediction of visual field deficits by diffusion tensor imaging in temporal lobe epilepsy surgery. *Neuroimage* **2009**, *45*, 286–297. [CrossRef] [PubMed]
23. Egan, R.A.; Shults, W.T.; So, N.; Burchiel, K.; Kellogg, J.X.; Salinsky, M. Visual field deficits in conventional anterior temporal lobectomy versus amygdalohippocampectomy. *Neurology* **2000**, *55*, 1818–1822. [CrossRef] [PubMed]
24. Guenot, M.; Krolak-Salmon, P.; Mertens, P.; Isnard, J.; Ryvlin, P.; Fischer, C.; Vighetto, A.; Mauguiere, F.; Sindou, M. MRI assessment of the anatomy of optic radiations after temporal lobe epilepsy surgery. *Stereotact. Funct. Neurosurg.* **1999**, *73*, 84–87. [CrossRef] [PubMed]
25. Hervas-Navidad, R.; Altuzarra-Corral, A.; Lucena-Martin, J.A.; Castaneda-Guerrero, M.; Vela-Yebra, R.; Sanchez, A. Defects in the visual field in resective surgery for temporal lobe epilepsy. *Rev. Neurol.* **2002**, *34*, 1025–1030. [PubMed]
26. McDonald, C.R.; Hagler, D.J., Jr.; Girard, H.M.; Pung, C.; Ahmadi, M.E.; Holland, D.; Patel, R.H.; Barba, D.; Tecoma, E.S.; Iragui, V.J.; et al. Changes in fiber tract integrity and visual fields after anterior temporal lobectomy. *Neurology* **2010**, *75*, 1631–1638. [CrossRef] [PubMed]
27. Mengesha, T.; Abu-Ata, M.; Haas, K.F.; Lavin, P.J.; Sun, D.A.; Konrad, P.E.; Pearson, M.; Wang, L.; Song, Y.; Abou-Khalil, B.W. Visual field defects after selective amygdalohippocampectomy and standard temporal lobectomy. *J. Neuroophthalmol.* **2009**, *29*, 208–213. [CrossRef] [PubMed]
28. Manji, H.; Plant, G.T. Epilepsy surgery, visual fields, and driving: A study of the visual field criteria for driving in patients after temporal lobe epilepsy surgery with a comparison of Goldmann and Esterman perimetry. *J. Neurol. Neurosurg. Psychiatry* **2000**, *68*, 80–82. [CrossRef] [PubMed]
29. Taoka, T.; Sakamoto, M.; Nakagawa, H.; Nakase, H.; Iwasaki, S.; Takayama, K.; Taoka, K.; Hoshida, T.; Sakaki, T.; Kichikawa, K. Diffusion tensor tractography of the Meyer loop in cases of temporal lobe resection for temporal lobe epilepsy: Correlation between postsurgical visual field defect and anterior limit of Meyer loop on tractography. *AJNR Am. J. Neuroradiol.* **2008**, *29*, 1329–1334. [CrossRef] [PubMed]
30. Winston, G.P.; Daga, P.; Stretton, J.; Modat, M.; Symms, M.R.; McEvoy, A.W.; Ourselin, S.; Duncan, J.S. Optic radiation tractography and vision in anterior temporal lobe resection. *Ann. Neurol.* **2012**, *71*, 334–341. [CrossRef] [PubMed]

31. Winston, G.P. Epilepsy surgery, vision, and driving: What has surgery taught us and could modern imaging reduce the risk of visual deficits? *Epilepsia* **2013**, *54*, 1877–1888. [CrossRef] [PubMed]
32. Wu, W.; Rigolo, L.; O'Donnell, L.J.; Norton, I.; Shriver, S.; Golby, A.J. Visual pathway study using in vivo diffusion tensor imaging tractography to complement classic anatomy. *Neurosurgery* **2012**, *70*, 145–156. [CrossRef] [PubMed]
33. Wang, Y.X.; Zhu, X.L.; Deng, M.; Siu, D.Y.; Leung, J.C.; Chan, Q.; Chan, D.T.; Mak, C.H.; Poon, W.S. The use of diffusion tensor tractography to measure the distance between the anterior tip of the Meyer loop and the temporal pole in a cohort from Southern China. *J. Neurosurg.* **2010**, *113*, 1144–1151. [CrossRef] [PubMed]

behavioral sciences

MDPI

Article

Oxidative Stress in Patients with Drug Resistant Partial Complex Seizure

Lourdes Lorigados Pedre [1], Juan M. Gallardo [2], Lilia M. Morales Chacón [3], Angélica Vega García [7], Monserrat Flores-Mendoza [7], Teresa Neri-Gómez [4], Bárbara Estupiñán Díaz [5], Rachel M. Cruz-Xenes [6], Nancy Pavón Fuentes [1] and Sandra Orozco-Suárez [7,*]

[1] Immunochemical Department, International Center for Neurological Restoration, 25th Ave, Playa, Havana 15805, Cuba; lourdesl@neuro.ciren.cu (L.L.P.); nancy@neuro.ciren.cu (N.P.F.)
[2] Medical Research Unit in Nephrological Diseases, Specialty Hospital, National Medical Center "XXI Century", IMSS, Mexico City 06720, Mexico; jmgallardom@gmail.com
[3] Clinical Neurophysiology Lab., International Center for Neurological Restoration, Havana 11300, Cuba; lily@neuro.ciren.cu
[4] Nanomaterials Laboratory, Research Center in Health Sciences, Autonomous University of San Luis Potosí, San Luis Potosi 78300; Mexico; tnerigomez@gmail.com
[5] Morphological Laboratory, International Center for Neurological Restoration, Havana 11300, Cuba; baby@neuro.ciren.cu
[6] Biology Faculty, University of Havana, Havana 10400, Cuba; rachel.cruz@fbio.uh.cu
[7] Medical Research Unit in Neurological Diseases, Specialty Hospital, National Medical Center, XXI Century, IMSS, Mexico City 06720, Mexico; ange_li_k@hotmail.com (A.V.G.); moonsefm@hotmail.com (M.F.-M.)
* Correspondence: sandra.orozcos@imss.gob.mx or sorozco5@hotmail.com; Tel.: +52-5555780240

Received: 18 April 2018; Accepted: 23 May 2018; Published: 9 June 2018

Abstract: Oxidative stress (OS) has been implicated as a pathophysiological mechanism of drug-resistant epilepsy, but little is known about the relationship between OS markers and clinical parameters, such as the number of drugs, age onset of seizure and frequency of seizures per month. The current study's aim was to evaluate several oxidative stress markers and antioxidants in 18 drug-resistant partial complex seizure (DRPCS) patients compared to a control group (age and sex matched), and the results were related to clinical variables. We examined malondialdehyde (MDA), advanced oxidation protein products (AOPP), advanced glycation end products (AGEs), nitric oxide (NO), uric acid, superoxide dismutase (SOD), glutathione, vitamin C, 4-hydroxy-2-nonenal (4-HNE) and nitrotyrosine (3-NT). All markers except 4-HNE and 3-NT were studied by spectrophotometry. The expressions of 4-HNE and 3-NT were evaluated by Western blot analysis. MDA levels in patients were significantly increased ($p \leq 0.0001$) while AOPP levels were similar to the control group. AGEs, NO and uric acid concentrations were significantly decreased ($p \leq 0.004$, $p \leq 0.005$, $p \leq 0.0001$, respectively). Expressions of 3-NT and 4-HNE were increased ($p \leq 0.005$) similarly to SOD activity ($p = 0.0001$), whereas vitamin C was considerably diminished ($p = 0.0001$). Glutathione levels were similar to the control group. There was a positive correlation between NO and MDA with the number of drugs. The expression of 3-NT was positively related with the frequency of seizures per month. There was a negative relationship between MDA and age at onset of seizures, as well as vitamin C with seizure frequency/month. We detected an imbalance in the redox state in patients with DRCPS, supporting oxidative stress as a relevant mechanism in this pathology. Thus, it is apparent that some oxidant and antioxidant parameters are closely linked with clinical variables.

Keywords: oxidative stress; drug-resistant epilepsy; redox; MDA; 4-HNE; 3-NT; AGEs; SOD; nitric oxide; vitamin C

1. Introduction

Oxidative stress (OS) is a biochemical state in which reactive oxygen species (ROS) are generated. Since the 1970s, OS been associated with diverse physiological and pathological conditions, including epilepsy. At high concentrations, ROS react readily with proteins, lipids, carbohydrates, and nucleic acids, often inducing irreversible functional alterations or even complete destruction. OS is defined as an imbalance between oxidants, nitrosative stress and antioxidants, which results in a relative or actual excess of oxidative species, and this leads to disruptions in signaling, redox control, and/or molecular damage. OS is also involved in acute and chronic central nervous system (CNS)injury and is a major factor in the pathogenesis of neuronal damage [1]. OS is also involved in acute and chronic CNS injury and is a major factor in the pathogenesis of neuronal damage [2].

Epilepsy is one of the most common and severe brain disorders in the world, affecting at least 50 million people worldwide. It is characterized by recurrent spontaneous seizures due to an imbalance between cerebral excitability and inhibition, with a tendency towards uncontrolled excitability. Approximately, 60–80% of patients with epilepsy can be controlled with antiepileptic drugs [3]. In more than 60% of all cases, seizures remit permanently. Nevertheless, a substantial proportion of patients (30%) do not respond to antiepileptic drug medication, despite administration in an optimally monitored regimen. Such cases are often loosely termed drug-resistant, intractable or pharmacoresistant [4]. The majority of these patients suffer from the focal form of epilepsy. The areas of epileptogenesis in these cases are usually characterized by cell loss [5,6].

Emerging works have indicated the involvement of redox imbalance in epileptogenesis [7]. Increased oxidant generation has been demonstrated to be induced in epilepsy by recurrent seizures with high levels of OS biomarkers and low antioxidant defenses present in epileptic subjects [8]. However, whether OS is a consequence, a causative factor, or both, in mechanisms involved in seizures is not clear [9,10].

Accumulating evidence supports the association between OS and seizures in seizure generation and in the mechanisms associated with refractoriness to drug therapy. Alterations in the antioxidant enzymes [11–14] and increases in the indicators of oxidative damage to biomolecules, such as malondialdehyde (MDA), protein carbonyls and 8-hydroxy-2-deoxyguanosine and activation of nicotinamide adenine dinucleotide phosphate oxidase have been reported [10,12,13]. Similarly, data from animal studies suggests that prolonged seizure activity might result in increased production of ROS. Further, the generation of nitric oxide and peroxynitrite has been shown to precede neuronal cell death in vulnerable brain regions [15,16].

Despite the alterations in the redox state described in epilepsy, little is known about how they behave jointly in complex partial seizures that are refractory to drugs in terms of parameters, such as the expression of 3-nitrotirosine (3-NT), 4 hydroxynoneal (4-HNE), levels of oxidants and antioxidants, and their relationships with clinical parameters, such as duration of epilepsy, frequency of seizures and number of drugs (poly or monotherapy).

Neuronal cells in the brain are highly sensitive to oxidative stress; therefore, the prolonged excitation of neurons during seizures can lead to injury resulting from biochemical alterations and specifically, to the role played by the oxidation state [17]. Excessive ROS generation can cause damage to neuronal cells, inducing cell death via either apoptotic or necrotic pathways [18]. Taking into account this information and based on the results described previously by our group, in which apoptotic and necrotic death were evidenced in temporal lobe epileptic patients [19,20], we proposed the evaluation of oxidative stress in patients who were suffering from drug-resistant partial complex seizure (DRPCS).

There have been few studies that have evaluated markers of oxidative stress in DRPCS in terms of parameters such as the expression of 3-nitrotirosine (3-NT), 4 hydroxynoneal (4-HNE), markers of damage to proteins, lipids and advanced glycation products (AGEs), and nitric oxide (NO) as well as antioxidants (superoxide dismutase (SOD), gluthatione, vitamin C and uric acid). The main objective of this work is not only to evaluate a large number of oxidative stress markers but also to explore the possible relationship of these markers with clinical parameters, such as the time of crisis evolution, the number of seizures per month and the medication received by these patients, e.g., monodrug or multidrug therapy.

2. Materials and Methods

2.1. Patient Information

The serum concentration of oxidative stress markers was studied in eighteen patients from the International Center for Neurological Restoration (CIREN) with DRCPS (6 females and 12 males, with mean age 33.28 ± 12.36 years). The criteria to be considered drug-resistant were the following: the presence of seizures for more than two years, two complex partial monthly seizures, two cycles of monotherapy and at least one cycle of polytherapy. All patients were evaluated in the Video-EEG Telemetry Unit from CIREN, and they underwent a complete general and neurological physical examination and anatomical evaluation by Magnetic Resonance Imaging (1.5 T MAGNETOM SINPHONY) and SPECT.

In terms of antiepileptic treatment, the most commonly used antiepileptic drugs (AEDs) by patients in the study were carbamazepine (3–6 tablets daily), followed by lamotrigine (2–3 tablets daily) and magnesium valproate (1–6 tablets daily).

To correlate OS markers with the number of medications taken by patients, treatments were grouped as follows: monodrug therapy (a single AED) or multidrug therapy (two or three AEDs).

The collected clinical parameters included demographic characteristics and clinical states, as shown in Table 1. The parameters were age, gender, time of seizure evolution, frequency of seizure by month and seizure localization and medication.

Table 1. Clinical data of patients with drug resistant complex partial seizures.

No.	Age (Years)	Gender	Time Seizure Evolution (Years)	Seizure Frequency/Month	Seizure Localization	Drugs
1	21	M	17	4,5	ET	PNT, PM, CBZ
2	19	M	16	16	ET	LMG, CBZ, CLON
3	48	F	35	13	ET	CBZ, LMG
4	19	M	17	2	T	TP
5	17	M	10	7	ET	CBZ, LMG, CLB
6	22	M	16	2	T	CBZ
7	57	M	11	12	T	MV
8	38	F	18	1	ET	MV
9	46	F	46	4	ET	PBT, PNT, CBZ
10	28	F	14	90	ET	CBZ, CLON
11	30	M	18	12	T	CBZ, LMG
12	38	M	30	3	T	CBZ
13	36	M	27	2	T	CBZ, MV
14	31	F	28	120	ET	LMG, CLON
15	27	M	10	1	T	MV
16	48	M	4	30	T	LMG
17	22	F	7	30	ET	CBZ, LMG
18	16	M	9	360	T	TOP, LVT, CLON

CBZ: carbamazepine, CLB: clobazam, CLON: clonazepam, ET: extratemporal, LMG: lamotrigine, LVT: levetiracetam, PM: primidone, PBT: phenobarbital, PNT: phenytoin, MV: magnesium valproate T: temporal, TOP: topiramate.

2.2. Samples

Venous blood samples (5 mL) were taken from the patients by antecubital puncture after asepsis of the region. The blood was centrifuged, and the serum was kept at −20 °C until its use. The serum samples were used to analyze the following routine biochemical parameters and oxidative biomarkers: lipid peroxidation measured by MDA, advanced oxidation protein products (AOPP), advanced glycation end products (AGEs); nitric oxide (NO), 3-NT, 4-HNE, uric acid, vitamin C and SOD.

Patients with chronic medical diseases (hypertension, diabetes, collagen vascular diseases, rheumatoid arthritis, other neurological diseases), those who were smokers, alcoholics, and those taking any other medications were excluded from the study.

The evaluation included analyses of seizure semiology, video-EEG monitoring using noninvasive methods, anatomical neuroimaging (MRI) and functional neuroimaging (SPECT, functional MRI) and neuropsychological assessment. None of the patients or controls received any antioxidant supplementation.

The control group was formed by 80 healthy volunteers who were aged 31.3 ± 8.64 including 50 males and 30 females matched to patients group. It was taken into account that the supposedly healthy subjects did not show antecedents of neurological diseases, were not smokers, alcoholics and did not take any medication or medical supplements.

2.3. Determination of Lipid Peroxidation

Lipid peroxidation was assessed by measuring MDA using the thiobarbituric acid (TBA)-reactive substances test [21] and was calculated in μmol/L. To an aliquot of plasma, 200 μL of 25% trichloroacetic acid was added. The samples were incubated at 4 °C for 15 min, followed by centrifugation at 4 °C, $5000 \times g$ for 3 min, and the supernatant (100 μL) was neutralized with 4 M NaOH (JT Baker, Xalostoc, Edo. De Mexico, Mexico). Then, to 1 mL of the above solution, 1 mL of 0.7% TBA (Acros Organics, Belgium) was added. This mixture was incubated at 90 °C for 60 min. The color reaction was measured spectrophotometrically (532 nm) in the organic phase (1-butanol, Sigma-Aldrich, St. Louis, MO, USA). Tetramethoxypropane was used as a standard. The results were expressed as TBARs per μmol/L.

2.4. AOPP Assay

AOPP were measured by spectrophotometry on a microplate reader (Multiskan, Thermolab) and were calibrated with chloramine-T solutions (Sigma, St. Louis, MO, USA) which, in the presence of potassium iodide, is absorbed at 340 nm [22]. Two hundred microliters of plasma diluted in a ratio of 1:5 in PBS was placed on a 96-well microtiter plate, and 20 μL of acetic acid was added. In standard wells, 10 μL of 1.16 M potassium iodide (Sigma, St. Louis, MO, USA) was added to 200 μL of chloramine-T solution (0–100 μmol/L) followed by 20 μL of acetic acid. The absorbance of the reaction mixture was immediately read at 340 nm on the microplate reader against a blank containing 200 μL of PBS, 10 μL of potassium iodide, and 20 μL of acetic acid. The chloramine-T absorbance at 340 nm was linear within the range of 0 to 100 μmol/L. AOPP concentrations were expressed as micromoles per liter of chloramine-T equivalent. AOPP were expressed in umol/L.

2.5. Advanced Glycation End Products AGEs

AGE concentrations were measured after the blood had been centrifuged. For fluorescent AGEs, we employed a 96-plate spectrophotofluorimeter (Fluoroskan Ascent FL. Vantaa, Finland). Briefly, 100 μL of serum was deproteinized with TCA (ReactivosQuimica Meyer, Mexico City, Mexico) at a concentration of 300 mmol/L. Then, 200 μL chloroform was added, vortexed for 60 s and centrifuged at 14,000 rpm. Finally, 200 μL of the respective supernatant was placed in each well, in triplicate fluorescence. The intensity was read at 440 nm after excitation at 355 nm. Results were expressed as arbitrary units (AU) corrected by serum proteins (measured by absorptiometry at 280 nm). The intra-assay variation coefficient was 8.4%. To ensure adequate readings by the spectrophotofluorimeter, fluorescence calibration curves were performed using quinine sulfate as the standard, which has similar excitation and emission spectra (360 and 440 nm, respectively).

2.6. Nitric Oxide

NO was measured by determining the total quantity of nitrite ($NO_2{}^-$), which is the stable product of NO metabolism in plasma. Griess reagent was used (an aqueous solution of 1% sulfanilamide (Sigma-Aldrich, St. Louis, MO, USA) with 0.1% naphthylethylenediamine (Sigma-Aldrich, St. Louis,

MO, USA) in 2.5% H_3PO_4 (2.5%, JT Baker, Xalostoc, Mexico), which forms a stable chromophore with NO_2^- and absorbs light at 546 nm (Green LC 1982). The calibration curve was constructed using different concentrations of sodium nitrite dissolved in 0.9% NaCl. The NO level was expressed in μmol/L.

2.7. Western Blot Analysis for 3-Nitrotyrosine and 4-Hidroxynonenal

The samples of serum were quantified according to the Bradford method (Bio-Rad, Hercules, CA, USA). Proteins (12 μg) were separated by electrophoresis on 10% SDS–PAGE gel at 80 V. Gels were transferred to Immun-Blot PVDF membrane for protein blotting (Bio-Rad) at 20 V at room temperature for 1 h. Membranes were blocked with blocking buffer (Millipore, Burlington, MA, USA) and incubated at 4 °C overnight with primary antibodies diluted in a ratio of 1:3000. Pre-stained broad range markers (BioRad, CA, USA) were included for size determination.

The following primary antibodies were used: goat anti-4-hidroxynonenal polyclonal antibody (AB5605 Millipore, Burlington, MA, USA), rabbit anti-3-NT polyclonal antibody, and rabbit anti-transferrin polyclonal antibody (Santa Cruz Biotechnology, CA, USA). After incubation with the primary antibody, membranes were washed and incubated with horseradish peroxidase-coupled secondary antibodies (VECTOR, diluted 1:15,000). Immunoreactive bands were detected by using an enhanced chemiluminescence system (Clarity Western ECL substrate by BIO-RAD). Membranes were stripped by employing a commercial solution (Millipore, Burlington, MA, USA) retested with anti-transferrin polyclonal antibody and detected by the system mentioned above. Anti-transferrin was used to correct the differences in the total amount of loaded protein. The intensity of the protein bands was quantified by densitometry using a molecular imager Fusion FX VilberLourmat (Vilber, Marne-la-Vallée, France), and captured data was quantified and analyzed by densitometry using Quantity One Image Analysis Software. The density of each band was normalized to its respective loading control (transferrin) and was expressed as the integrated density value.

2.8. SOD Activity

SOD was determined by the Marklund and Marklund method [23]. The anionic superoxide radical participates in the autoxidation of pyrogallol. To this end, a pyrogallol solution was prepared in HC1 (JT Baker, Xalostoc, Mexico) and incubated at 40 °C. To 50 μL of sample, 200 μL of a mixture of Tris-EDTA-HC1 was added and read at 420 nm in a DU 50 spectrophotometer (Beckman, Palo Alto, CA, USA). Subsequently, the pyrogallol solution was added and the increase in absorbance was re-measured every 30 s for 3 min. The reagent blank was made in the same way but distilled water was used instead of sample. The activity of SOD is expressed in U/gHb.

2.9. Vitamin C

The measurement of vitamin C was done according Prieto et al. [24]. This method is based on the reduction of molybdate (VI) to molybdate (V) by the sample, followed by the formation of a complex between phosphate and molybdate (V), which has an optimal absorption at 695 nm. The calibration curve was prepared with ascorbic acid and read at 695 nm. The vitamin C concentration is expressed in mmol/L.

2.10. Uric Acid

Uric acid is a metabolite of purines, nucleic acids and nucleoproteins. Usually, the concentration of uric acid in serum varies from one individual to another according to various factors, such as sex, diet, ethnicity, genetic makeup, pregnancy, and many other conditions. Uric acid was measured using the enzymatic Uricostat reagent kit (Wiener brand, Rosario, Argentina), following the manufacturer's instructions. The levels of uric acid are expressed in mg/dL.

2.11. Gluthatione

The activity of glutathione was analyzed using the method of Beutler, et al. [25]. Gluthatione was measured with 5,5 'dithiobis- (2-nitrobenzoic acid) (DTNB). The reaction mixture contained 1 mL of gluthatione (Amresco), 2 mmol of buffer, 400 mmol of PBS (pH 7.0), 4 mmol of EDTA, 1 mmol of 0.5% sodium azide, 250 μL of seminal fluid or sperm and bidistilled water to give a total of 4 mL. After incubation at 37 °C for five minutes, 1 mL of prewarmed T-BOOH at a concentration of 1.25 mmol was added and re-incubated for four more minutes. At the end of that period, 1 mL was recovered, and 4 mL of phosphoric acid was added, centrifuged at 2000× g, at room temperature, for 10 min. Two milliliters of the supernatant was recovered, and 2 mL of Na_2HPO_4 400 mmol and 1 mL of the DTNB reagent were added. The absorbance was measured at 412 nm. The targets and standards were prepared similarly. The activity of glutathione was expressed as U/mg of protein.

2.12. Ethical Considerations

All procedures followed the rules of the Declaration of Helsinki of 1975 for human research, and the study was approved by the Ethics Committee of the International Center for Neurological Restoration (Record 03/2015).

2.13. Statistical Processing

Statistical analysis was carried out using GraphPad Prism 5 software (GraphPad Software, Inc., La Jolla, CA, USA). The values are expressed as means ± SEMs. Normal distribution and homogeneity of variance of the data were tested with the Kolmogorov–Smirnov and Levene tests, respectively. The comparisons between two groups were made by means of the t-Student test. The Spearman correlation was used for the correlation study. In all cases, statistically significant differences were considered when $p \leq 0.05$.

3. Results

There has been previous evidence supporting a gender difference in epilepsy related to OS markers [26,27]. In this study there were no gender differences for all OS markers evaluated, so the concentrations in males and females were unified into one group for each marker.

3.1. Proteins and Lipid Damage and Advanced Glycation

The brain has a high lipid content, and its oxygen consumption and oxidative metabolism make it susceptible to oxidative stress. Lipid peroxidation involves the oxidative degradation of lipids. Protein oxidation is irreversible oxidative damage, and protein damage is considered to be a marker for severe oxidative stress. In order to evaluate the damage to proteins and lipids, we measured the concentrations of MDA, AOPP and AGEs in DRPCS patients and compared these with the control group (Figure 1). The results showed a statistically significant increase in MDA ($p \leq 0.00001$, Figure 1A) in patients (39.78 ± 3.23 μm/L) when compared with controls (18.23 ± 0.81 μm/L). No differences were found between groups with regard to AOPP; however, it is possible to observe a trend towards the steric increment in these products (patients: 48.04 ± 3.61 μm/L, controls: 42.73 ± 2.53 μm/L Figure 1B). There was a significant decrease in AGEs ($p \leq 0.0049$, Figure 1C) in patients (2.38 ± 0.21 UAF) versus controls (2.82 ± 0.04 UAF).

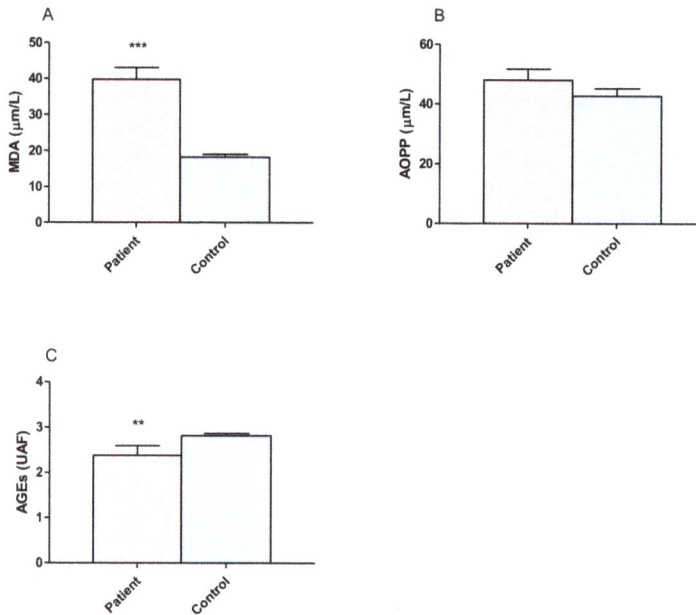

Figure 1. Comparison of concentrations of malondialdehyde (MDA), advanced oxidation protein products (AOPP) and advanced glycation end products (AGEs) between patients with drug resistant partial complex seizure and controls. (**A**) MDA serum concentration. (**B**) AOPP serum concentration. (**C**) AGEs serum concentration. The bars represent the means ± the standard errors of the mean. Unpaired t-Student test. ** $p \leq 0.004$, *** $p \leq 0.0001$.

3.2. Nitric Oxide

Nitric oxide is a free radical that is formed biologically through the oxidation of L-arginine by nitric oxide synthase. The NO levels in patients (9.85 ± 2.34 μm/L) were lower than those of the control group (25.31 ± 2.14 μm/L), with a significant level of $p \leq 0.0054$ (Figure 2).

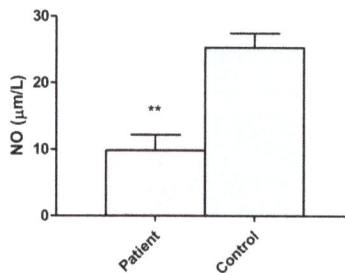

Figure 2. Comparison of nitric oxide (NO) concentrations between patients with drug resistant partial complex seizure and controls. The bars represent the means ± standard errors of the mean. Unpaired t-Student test. ** $p \leq 0.0054$.

3.3. Expression of 4-Hydroxy-2-Noneal and 3-Nitrotyrosine

3-NT is an oxidative marker of NO inactivation, which, in our results, showed a statistically significant increase in the expression of 3-NT in group of patients with DRCPS (18.40 ± 3.39) compared with the control subjects (2.65 ± 0.79, Figure 3A).

The expression of 4-HNE (one of the major end products of lipid peroxidation) in patients (40.04 ± 5.53) was increased ($p \leq 0.00001$) compared with the control group (2.52 ± 0.39, Figure 3B).

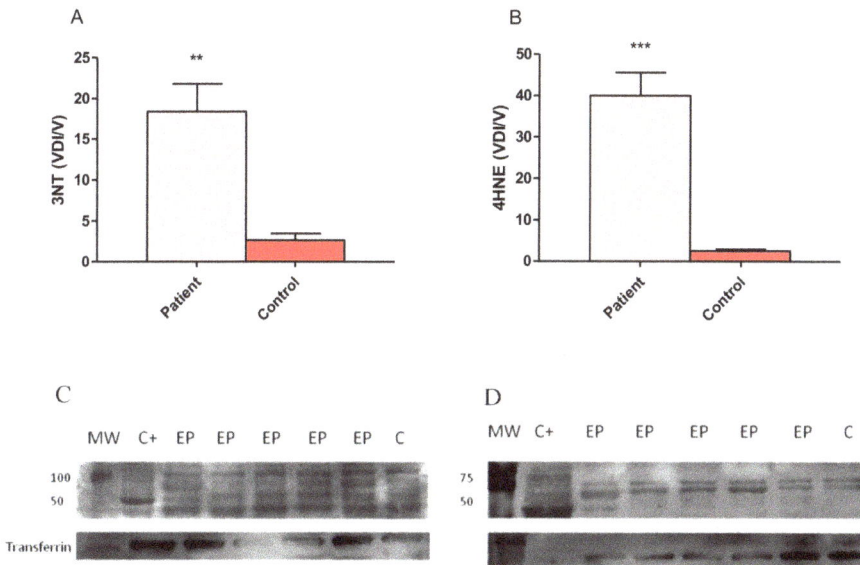

Figure 3. 3-Nitrotyrosine (3-NT) and 4-hydroxy-2-nonenal (4-HNE) expression in patients with drug resistant partial complex seizure. (**A**) Representative example of the blot of 3-NT in a patient and transferrin as a control protein. (**B**) Representative example of the blot of 4-HNE in a patient and transferrin and controls. (**C**) Comparison of relative optical density values of 3-NT in patients and control subjects. (**D**) Comparison of relative optical density values of 4-HNE in patients and control subjects. The bars represent the means ± the standard errors of the mean, unpaired t-Student test, ** $p \leq 0.003$, *** $p \leq 0.0001$. C+ rat protein extract was used as the positive control in both blots. MW; molecular weight, EP; epileptic patient, C; control.

3.4. Antioxidant Evaluation

In order to evaluate the serum concentrations of antioxidants in DRPCS patients, we measured the concentrations of SOD, glutathione, vitamin C and uric acid. The results showed a statistically significant increase in SOD ($p \leq 0.00001$, Figure 4A). There were no differences between the glutathione levels in patients versus the control group ($p \leq 0.07$); however, a tendency to decrease this parameter in patients was observed (Figure 4B). Significantly lower values of vitamin C and uric acid were shown in patients (3.33 ± 0.42 mm/L, 6.51 ± 0.05 mg/dL, Figure 4C,D respectively.) when compared with control group (34.72 ± 1.32 mm/L, 6.51 ± 0.05 mg/dL respectively).

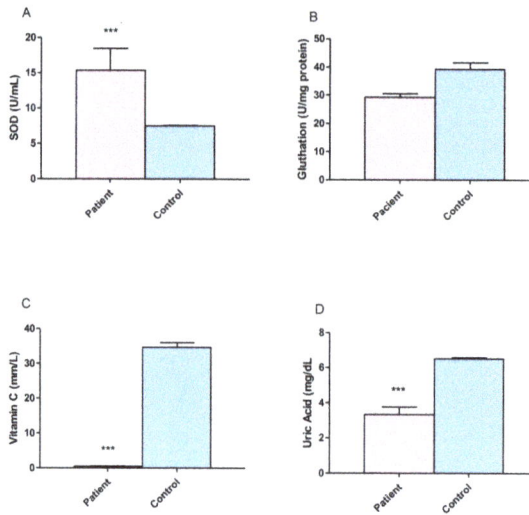

Figure 4. Comparison of superoxide dismutase (SOD), glutathione, vitamin C and uric acid concentrations between patients with drug resistant partial complex seizure and controls. (**A**) SOD serum concentration. (**B**) Gluthatione serum concentration. (**C**) Vitamin C serum concentration and (**D**) Uric acid serum concentration. The bars represent the means ± standard errors of the mean. Unpaired t-Student test. *** $p \leq 0.0001$.

3.5. Correlation of Oxidative Stress Parameters with Clinical Data

Table 2 shows the correlation between the parameters of oxidative stress evaluated in this study with clinical data such as the time evolution of seizures, the frequency of seizures per month and the number of drugs taken by a patient. There was a positive correlation between MDA and NO with the number of drugs. Patients with multidrug treatments had the highest levels of MDA and NO, while the elevation of MDA was related to the early age of onset of seizures. Similarly, the frequency of crises per month were negatively related to vitamin C and positively related to the expression of 3-NT.

Table 2. Correlation of MDA, NO, Vvtamin C and 3-NT with the time evolution of seizures, the frequency of seizures per month and the number of drugs taken by a patient.

Oxidative Stress Parameters	Age at Onset of Seizure	Frequency of Seizures per Month	Number of Drugs
MDA	−0.5500 *	0.0484	0.5351 *
AOPP	−0.1050	0.1048	−0.0291
AGEs	−0.4741	−0.3939	0.3028
NO	−0.3416	−0.3818	0.5843 *
Uric Acid	−0.1853	−0.1996	0.0529
SOD	0.1503	−0.2642	−0.2594
Glutathione	−0.1652	−0.2892	0.2182
Vitamin C	−0.2136	−0.7110 *	0.1912
3−NT	0.3878	0.7565 *	−0.0670
4−HNE	−0.0852	0.3246	0.4022

Values represent the Spearman correlation coefficients. * $p \leq 0.05$; 3-NT: nitrotyrosine, 4-HNE: 4-hydroxy-2-nonenal, AGEs: advanced glycation end products, AOPP: advanced oxidation protein products, MDA: malondialdehyde, NO: nitric oxide, SOD: superoxide dismutase.

4. Discussion

The molecular mechanisms that lead to seizures and epilepsy are not well understood. Previous studies have demonstrated that seizure-induced mitochondrial dysfunction and excess free radical production cause oxidative damage to cellular components and initiate the mitochondrial apoptotic pathway [28,29]. OS is also considered an important consequence of excitotoxicity and inflammation, two of the proposed mechanisms for seizure-induced brain damage [20,29–34].

Previous studies have found increased activities of SOD, CAT, markers of lipid peroxidation and decreased activities of glutathione peroxidase (GPx) in pharmacoresistant temporal lobe epilepsy (TLE) patients [35,36]. Sudha K, (2001) reported a decrease in glutathione reductase. The lipid peroxidation and percentage hemolysis were higher compared to controls. Furthermore, erythrocyte glutathione reductase and plasma ascorbate and vitamin A concentrations were lower [8]. Meanwhile, many different studies have noted increased markers in lipid peroxidation [12,36]. On the other hand, there have also been other studies that have not detected changes in SOD, CAT, GPx and glutathione reductase activities [8,36–38].

Our group has been studying the impact of epilepsy surgery on serum markers of oxidative damage in pharmacoresistant temporal lobe epilepsy (TLE) patients [12]. Before surgery, we found increased activities of SOD, CAT, markers of lipid peroxidation and decreased GPx activity. An interesting finding was the positive correlation between the duration of the disease and advanced oxidation protein product levels. This result suggests the early presence of oxidative damage to proteins in the initial stages of the illness. This could be due to protein repairing mechanisms that do not act as efficiently as in other biomolecules. After surgery, the patients showed a tendency for the studied variables to normalize, except for SOD activity. The outlying redox state of the patients markedly improved after surgery, which was clearly evidenced by decreases in MDA and advanced oxidative protein products levels two years after surgery. The recovery in GPx activity was also notorious, as it contributed to a decrease in oxidative damage and a better redox balance [12]. On the other hand, we can speculate that the sustained increase in superoxide dismutase activity could recede if the epileptic activity in the remaining regions eventually disappears in these patients. Finally, the increase in CAT activity levels seems to be a cellular response to the intense ROS production triggered by seizure episodes.

On the other hand, there is information that supports the suggestion that inflammation and oxidative stress are linked with a number of chronic diseases, including diabetes and diabetic complications, hypertension, cardiovascular diseases, neurodegenerative diseases, and aging [39–42]. Inflammatory cells liberate a number of reactive species at the site of inflammation, leading to exaggerated oxidative stress [43,44]. Secondly, a number of reactive oxygen/ nitrogen species can initiate intracellular signaling cascade that enhances proinflammatory gene expression [45]. There is a group of previous findings from our working group that supports the participation of inflammatory processes in drug-resistant epilepsy [46], which could be a source of the OS detected in these patients. Thus, inflammation and oxidative stress are closely related pathophysiological events that are tightly linked with one another.

4.1. Proteins, Lipid Damage and Advanced Glycation Products

Lipid peroxidation is one of the major sources of free radical-mediated injury that directly damages membranes and generates a number of secondary products. In particular, markers of lipid peroxidation have been found to be elevated in brain tissues and body fluids in several neurodegenerative diseases, such as epilepsy. This complex process involves the interaction of oxygen-derived free radicals with polyunsaturated fatty acids (PUFAs), resulting in a variety of highly reactive electrophilic aldehydes, including MDA, 4-HNE, and acrolein. Therefore, the evaluation of MDA levels in biological materials can be used as an important indicator of lipid peroxidation for various diseases. In the present study, we found significantly higher levels of MDA in patients with epilepsy compared to controls; these results are in agreement with previous studies in which it has been reported that a higher level of MDA is associated with epilepsy [47,48]. On the other hand, these studies showed that the levels of MDA were significantly increased in treated patients (with AED) compared to untreated patients, which was also observed in patients with epilepsy. There was a significant correlation with the number of drugs administered (Table 2) which suggests that additional oxidative stress was induced by AEDs, causing the recurrence of seizures and intolerance to drugs. Peroxidation of membrane lipids can have numerous effects, such as increased membrane rigidity, decreased activity of membrane-bound enzymes (e.g., sodium pump), altered activity of membrane receptors, and altered permeability [49]. In addition to effects on phospholipids, radicals can also directly attack membrane proteins and induce lipid–lipid, lipid–protein, and protein–protein cross-linking, all of which obviously have effects on membrane function [50]. Of these products, MDA, 4-HNE, and acrolein can cause irreversible modification of phospholipids, proteins, and DNA, resulting in impaired function and consequently, cell death, a fact observed in complex partial epilepsy, which has been demonstrated in previous works involving pharmacoresistant epilepsy patients [19,46]. On the other hand, the formation of AGEs, a group of modified proteins and/or lipids with damaging potential, is one contributing factor. However, it has been reported that AGEs increase reactive oxygen species formation and impair antioxidant systems, while the formation of some AGEs is induced under oxidative conditions. However, in the patients with epilepsy, the values were lower than those in controls; thus, AGEs contribute less to chronic stress conditions in epilepsy.

Lipid peroxidation alters membrane structure, affecting its fluidity and permeability and the activity of membrane-bound proteins and produces many cytotoxic and reactive by-products. Among these, 4-HNE is able to form adducts with biomolecules, including proteins, lipids and nucleic acids, thereby propagating oxidative damage [51]. HNE-mediated damage to proteins is a known oxidative posttranslational modification that leads to functional changes or deactivation of enzymes, transporters, ion channels and receptors [52].

In this study, the levels of damage to biomolecules increased and coincidentally, the expression of 4-HNE is high in the patients in relation to the control group. Furthermore, an increase in 4-HNE occurs in various pathological conditions, including neurological diseases, where it contributes to cell death and neurodegeneration [52,53]].

4-HNE is considered to be a "second toxic messenger" that can propagate and amplify initial oxidative injury. 4-HNE can form covalent bonds with three different amino acyl side chains, i.e., lysyl, histidyl, and cysteinyl residues. In addition, 4-HNE can modify protein structure through Schiff base formation with lysyl residues, leading to the formation of pyrrole, and/or form intra- and/or intermolecular cross-links. Due to its amphiphilic nature, hydroxyaldehyde can diffuse across membranes and covalently modify proteins in the cytoplasm and nucleus, far from their site of origin [52].

Our results support an increase in the expression of 4-NHE in patients with DRPCS, which coincides with that reported by other authors, such as Pecorelli et al., that describe high levels of 4-HNE-protein adduct brain tissues and body fluids in several neurological and neurodegenerative diseases and in drug-resistant epileptic patients [10,54,55]. This confirms the evidence of lipid peroxidation and indicates the presence of oxidative damage to proteins in human epileptic brain. Other studies have evaluated 4-HNE levels in epileptic diseases and reported high levels of brain 4-HNE during seizures in the kindling model [56]. In addition, it is well known that HNE can be an apoptotic inducer [57]. High levels of 4-HNE may eventually promote cell death; there is evidence that 4-HNE plays a pivotal role in neuronal death, as has already been demonstrated in several neurological diseases [53,58,59].

NO is a free radical that is formed biologically through the oxidation of L-arginine by nitric oxide synthase. The exact role of NO in the pathophysiology of epilepsy is still unclear. Several studies have shown that NO may act as an endogenous anticonvulsant [60–62], although some studies have shown that NO acts as a proconvulsant [63,64]. Our results showed a significant decrease in NO in patients compared with the control group and a relationship with the number of AEDs that patients receive.

On the other hand, 3-NT is an indicator for protein nitration, a posttranslational modification specific to the tyrosine amino acid which can yield protein dysfunction or turnover. A primary source and major contributor to tyrosine nitration in physiological and pathological events in vivo is through

ONOO- production, a reaction by-product of NO and O2•− [65]. We described an increase in the expression of 3-NT. Similarly, other authors have found an increase in 3-NT, but in an experimental model study using pilocarpine [66]. They specifically described the fact that 3-NT accumulates in hippocampal CA3, CA1 and hilar neurons following kainate-induced status epilepticus [66].

4.2. Antioxidants

SOD plays a crucial role in the elimination of superoxide anion radicals (O2•−) generated from extracellular stimulants, which include ionizing radiation and oxidative insults, along with those produced in the mitochondrial matrix. In the present study, we found a significant increase in SOD activity in patients with epilepsy compared to controls. However, this was not so for glutathione, uric acid and vitamin C, that were reduced as compared with control group; these results are in agreement with previous studies, where a significantly lower activity has been found in patients with epilepsy [47].

Previous studies on SOD activity have mainly been performed in children with epilepsy on different AEDs, but they reported no significant differences in SOD activity between children with epilepsy and controls [37,67,68]. Recently, different results have shown that, after a seizure, it is possible to observe an increase of SOD activity and there is a significant decrease in SOD activity only in patients receiving lamotrigine, while valproic acid and carbamazepine were associated with weaker effects. Furthermore, after 6 months of treatment with AEDs, the activity of this antioxidant enzyme remained significantly higher in patients than in individuals in the control group [37].

The incremental SOD activity in the epileptic group subjects suggests efficient conversion of superoxide radicals to less toxic H_2O_2, and a reduction in the accumulation of superoxide radicals, and a compensatory mechanism reduced GPx activity or decreased the antioxidant activities of GPx and vitamin C and increased the activity of MDA, suggesting a reduced antioxidant defense capacity in epilepsy patients, and not one ameliorated by antiepileptic treatment.

4.3. Correlation of Oxidative Stress Parameters with Clinical Data

The results of the correlation analysis of the clinical parameters with the oxidative stress markers showed a negative relationship of MDA with the age of onset of the crises and a positive relationship with the number of drugs taken (Table 2) as the level of MDA significantly increased. The correlation with a larger number of drugs suggests that the drugs induce additional oxidative stress, while it is known that long-term use of AEDs leads to the impairment of the endogenous antioxidant system. AEDs, namely valproic acid, phenytoin, CBZ, and levetiracetam, are shown to increase lipid peroxidation and decrease the activity of the antioxidant system [69]. Similarly, a positive correlation was found with NO levels and the number of drugs. NO plays a significant role in epilepsy and epileptogenesis, since it acts as a secondary messenger, neuromodulator and neurotransmitter [70]. It has been reported that significantly elevated levels of NO are associated with the severity of epilepsy [68]. However, in the present work, the serum levels were correlated with the number of drugs used, although the NO levels were lower in epilepsy patients than in the controls (Figure 3).

The marker of oxidative stress that was most strongly associated with the severity of the epilepsy was 3-NT, which is considered as a marker of NO-dependent oxidative stress, indicating that oxidative stress is induced by seizures and AEDs, which results in seizure recurrence and drug intractability. The antioxidant system that was most negatively correlated with the severity of epilepsy was vitamin C which was associated with a greater number of seizures. Very low levels of serum vitamin C were detected; low antioxidant biomolecule levels suggest a reduced antioxidant defense capacity in epilepsy patients.

An interesting result from this study is the relationship between some oxidants and multidrug therapy. There are different studies that have related the presence of OS with anticonvulsant treatment [68]. Some of these studies were conducted on patients who were already on treatment and thus, could not conclude that the oxidative stress was a result of epilepsy or AEDs. Menon [71] compared treated and untreated patients and found that the role of AEDs in increasing oxidative stress is negligible [71]. Nevertheless, our results support that the high values of damage to lipids evaluated by MDA are closely related to the use of two or more AEDs as therapy in these patients. However, our study had a relatively small number of subjects. Studies with larger sample sizes are needed to confirm these results.

5. Conclusions

In summary, our results indicate altered antioxidative defenses and damage to biomolecules in patients with DRPCS. The number of AEDs influences some oxidative markers, as do the age of onset of the crisis and the frequency of seizures. The mechanism of epileptogenesis is still unclear. In spite of the advent of newer AEDs, a significant proportion of patients are refractory to treatment. Whether free radicals released after seizure episodes lead to further seizures or refractory epilepsy needs to be addressed. Considering this knowledge, in the future, it would be interesting to correlate the oxidative markers with levels of proinflammatory mediators in order to elucidate the relationship between oxidative imbalance described for these patients and these mechanisms. This study adds information to the existing literature; however, further detailed research is needed to understand the mechanism of drugs resistant epilepsy, to promote newer modalities of disease modification.

Author Contributions: L.L.P. conceived and designed the experiments, produced and analyzed the results, and wrote the paper. J.M.G. executed the stress oxidative studies and the analysis of the results, L.M.M.C. evaluated the patients and participated in the writing of the paper, A.V.G. executed the Western blot study, M.F.-M. executed the Western blot study, T.N.-G. collaborated in the biochemistry studies, B.E.D. obtained and processed brain tissue. N.P.F., R.M.C.-X. participated in obtaining the samples and the analysis of result. S.O.-S. were advisors for the research and participated in the writing of the article.

Funding: This research received no external funding.

Acknowledgments: The authors would like to thank all the members of the epilepsy surgery program from the International Center for Neurological Restoration in Havana, Cuba and from the laboratory of Sandra and Gallardo, where the laboratory studies were carried out.

Conflicts of Interest: The authors declare no conflict of interest.

References

1. Cardenas-Rodriguez, N.; Huerta-Gertrudis, B.; Rivera-Espinosa, L.; Montesinos-Correa, H.; Bandala, C.; Carmona-Aparicio, L.; Coballase-Urrutia, E. Role of oxidative stress in refractory epilepsy: Evidence in patients and experimental models. *Int. J. Mol. Sci.* **2013**, *14*, 1455–1476. [CrossRef] [PubMed]
2. Uttara, B.; Singh, A.V.; Zamboni, P.; Mahajan, R.T. Oxidative stress and neurodegenerative diseases: A review of upstream and downstream antioxidant therapeutic options. *Curr. Neuropharmacol.* **2009**, *7*, 65–74. [CrossRef] [PubMed]
3. Hauser, W.A.; Annegers, J.F.; Kurland, L.T. Prevalence of epilepsy in rochester, minnesota: 1940–1980. *Epilepsia* **1991**, *32*, 429–445. [CrossRef] [PubMed]
4. Regesta, G.; Tanganelli, P. Clinical aspects and biological bases of drug resistant epilepsies. *Epilepsy Res.* **1999**, *34*, 109–122. [CrossRef]
5. Henshall, D.C.; Meldrum, B.S.; Soman, S.; Korah, P.K.; Jayanarayanan, S.; Mathew, J.; Paulose, C.S. Cell death and survival mechanisms after single and repeated brief seizures oxidative stress induced nmda receptor alteration leads to spatial memory deficits in temporal lobe epilepsy: Ameliorative effects of withania somnifera and withanolide a. *Neurochem. Res.* **2012**, *37*, 1915–1927.
6. Fujikawa, D.G.; Shinmei, S.S.; Cai, B. Kainic acid-induced seizures produce necrotic, not apoptotic, neurons with internucleosomal DNA cleavage: Implications for programmed cell death mechanisms. *Neuroscience* **2000**, *98*, 41–53. [CrossRef]
7. Aguiar, C.C.; Almeida, A.B.; Araujo, P.V.; de Abreu, R.N.; Chaves, E.M.; do Vale, O.C.; Macedo, D.S.; Woods, D.J.; Fonteles, M.M.; Vasconcelos, S.M. Oxidative stress and epilepsy: Literature review. *Oxid. Med. Cell. Longev.* **2012**, *2012*, 1–12. [CrossRef] [PubMed]

8. Sudha, K.; Rao, A.V.; Rao, A. Oxidative stress and antioxidants in epilepsy. *Clin. Chim. Acta* **2001**, *303*, 19–24. [CrossRef]
9. Waldbaum, S.; Patel, M. Mitochondrial dysfunction and oxidative stress: A contributing link to acquired epilepsy? *J. Bioenerg. Biomembr.* **2010**, *42*, 449–455. [CrossRef] [PubMed]
10. Pecorelli, A.; Natrella, F.; Belmonte, G.; Miracco, C.; Cervellati, F.; Ciccoli, L.; Mariottini, A.; Rocchi, R.; Vatti, G.; Bua, A.; et al. Nadph oxidase activation and 4-hydroxy-2-nonenal/aquaporin-4 adducts as possible new players in oxidative neuronal damage presents in drug-resistant epilepsy. *Biochim. Biophys. Acta* **2015**, *1852*, 507–519. [CrossRef] [PubMed]
11. Ben-Menachem, E.; Kyllerman, M.; Marklund, S. Superoxide dismutase and glutathione peroxidase function in progressive myoclonus epilepsies. *Epilepsy Res.* **2000**, *40*, 33–39. [CrossRef]
12. Lopez, J.; Gonzalez, M.E.; Lorigados, L.; Morales, L.; Riveron, G.; Bauza, J.Y. Oxidative stress markers in surgically treated patients with refractory epilepsy. *Clin. Biochem.* **2007**, *40*, 292–298. [CrossRef] [PubMed]
13. Ho, Y.H.; Lin, Y.T.; Wu, C.W.; Chao, Y.M.; Chang, A.Y.; Chan, J.Y. Peripheral inflammation increases seizure susceptibility via the induction of neuroinflammation and oxidative stress in the hippocampus. *J. Biomed. Sci.* **2015**, *22*, 46. [CrossRef] [PubMed]
14. Pearson-Smith, J.N.; Patel, M. Metabolic dysfunction and oxidative stress in epilepsy. *Int. J. Mol. Sci.* **2017**, *18*, 2365. [CrossRef] [PubMed]
15. Chuang, Y.C.; Chen, S.D.; Lin, T.K.; Liou, C.W.; Chang, W.N.; Chan, S.H.; Chang, A.Y. Upregulation of nitric oxide synthase ii contributes to apoptotic cell death in the hippocampal ca3 subfield via a cytochrome c/caspase-3 signaling cascade following induction of experimental temporal lobe status epilepticus in the rat. *Neuropharmacology* **2007**, *52*, 1263–1273. [CrossRef] [PubMed]
16. Chuang, Y.C.; Chen, S.D.; Liou, C.W.; Lin, T.K.; Chang, W.N.; Chan, S.H.; Chang, A.Y. Contribution of nitric oxide, superoxide anion, and peroxynitrite to activation of mitochondrial apoptotic signaling in hippocampal ca3 subfield following experimental temporal lobe status epilepticus. *Epilepsia* **2009**, *50*, 731–746. [CrossRef] [PubMed]
17. Mendez-Armenta, M.; Nava-Ruiz, C.; Juarez-Rebollar, D.; Rodriguez-Martinez, E.; Gomez, P.Y. Oxidative stress associated with neuronal apoptosis in experimental models of epilepsy. *Oxid. Med. Cell. Longev.* **2014**, *2014*, 293689. [CrossRef] [PubMed]
18. Shin, E.J.; Jeong, J.H.; Chung, Y.H.; Kim, W.K.; Ko, K.H.; Bach, J.H.; Hong, J.S.; Yoneda, Y.; Kim, H.C. Role of oxidative stress in epileptic seizures. *Neurochem. Int.* **2011**, *59*, 122–137. [CrossRef] [PubMed]
19. Lorigados, P.L.; Orozco Suárez, S.; Morales, C.L.; Garcia, M.I.; Estupinan, D.B.; Bender del Busto, J.E.; Pavon, F.N.; Paula, P.B.; Rocha, A.L. Neuronal death in the neocortex of drug resistant temporal lobe epilepsy patients. *Neurologia* **2008**, *23*, 555–565.
20. Lorigados, L.; Orozco Suárez, S.; Morales-Chacon, L.; Estupiñán, B.; García, I.; Rocha, L. Excitotoxity and neuronal death in epilepsy. *Biotecnol. Apl.* **2013**, *30*, 9–16.
21. Wade, C.R.; van Rij, A.M. Plasma thiobarbituric acid reactivity: Reaction conditions and the role of iron, antioxidants and lipid peroxy radicals on the quantitation of plasma lipid peroxides. *Life Sci.* **1988**, *43*, 1085–1093. [CrossRef]
22. Witko, V.; Nguyen, A.T.; Descamps-Latscha, B. Microtiter plate assay for phagocyte-derived taurine-chloramines. *J. Clin. Lab. Anal.* **1992**, *6*, 47–53. [CrossRef] [PubMed]
23. Marklund, S.L. Analysis of extracellular superoxide dismutase in tissue homogenates and extracellular fluids. *Methods Enzymol.* **1990**, *186*, 260–265. [PubMed]
24. Prieto, P.; Pineda, M.; Aguilar, M. Spectrophotometric quantitation of antioxidant capacity through the formation of a phosphomolybdenum complex: Specific application to the determination of vitamin e. *Anal. Biochem.* **1999**, *269*, 337–341. [CrossRef] [PubMed]
25. Beutler, E.; Duron, O.; Kelly, B.M. Improved method for the determination of blood glutathione. *J. Lab. Clin. Med.* **1963**, *61*, 882–888. [PubMed]
26. Carlson, C.; Dugan, P.; Kirsch, H.E.; Friedman, D. Sex differences in seizure types and symptoms. *Epilepsy Behav.* **2014**, *41*, 103–108. [CrossRef] [PubMed]
27. Brunelli, E.; Domanico, F.; La Russa, D.; Pellegrino, D. Sex differences in oxidative stress biomarkers. *Curr. Drug Targets* **2014**, *15*, 811–815. [CrossRef] [PubMed]
28. Henshall, D.C. Apoptosis signalling pathways in seizure-induced neuronal death and epilepsy. *Biochem. Soc. Trans.* **2007**, *35*, 421–423. [CrossRef] [PubMed]

29. Vezzani, A.; Friedman, A.; Dingledine, R.J. The role of inflammation in epileptogenesis. *Neuropharmacology* **2013**, *69*, 16–24. [CrossRef] [PubMed]
30. Vezzani, A.; Maroso, M.; Balosso, S.; Sanchez, M.A.; Bartfai, T. Il-1 receptor/toll-like receptor signaling in infection, inflammation, stress and neurodegeneration couples hyperexcitability and seizures. *Brain Behav. Immun.* **2011**, *25*, 1281–1289. [CrossRef] [PubMed]
31. Vezzani, A.; Balosso, S.; Ravizza, T. Inflammation and epilepsy. *Handb. Clin. Neurol.* **2012**, *107*, 163–175. [PubMed]
32. Castellanos Ortega, M.R.; Cruz, A.R.; Lorigados, P.L.; De La Cuetara, B.K. Purification and characterization of murine beta-nerve growth factor. *J. Chromatogr. B Biomed. Sci. Appl.* **2001**, *753*, 245–252. [CrossRef]
33. Lorigados, P.L.; Morales Chacon, L.M.; Orozco, S.S.; Pavon, F.N.; Estupinan, D.B.; Serrano, S.T.; Garcia, M.I.; Rocha, A.L. Inflammatory mediators in epilepsy. *Curr. Pharm. Des.* **2013**, *19*, 6766–6772. [CrossRef]
34. Ureña-Guerrero, M.E.; Feria-Velasco, A.; Gudiño-Cabrera, G.; Camin-Espuny, A.; Beas-Zrate, C. Modifications in the seizures susceptibility by excitotoxic neuronal damage and possible relationship with the pharmacoresistance. In *Pharmacoresistance in Eplepsy. From Genes and Molecules to Promising Therapy*; Rocha, A.L., Cavalheiro, E.A., Eds.; Springer: New York, NY, USA, 2013; pp. 59–76.
35. Yis, U.; Seckin, E.; Kurul, S.H.; Kuralay, F.; Dirik, E. Effects of epilepsy and valproic acid on oxidant status in children with idiopathic epilepsy. *Epilepsy Res.* **2009**, *84*, 232–237. [CrossRef] [PubMed]
36. Turkdogan, D.; Toplan, S.; Karakoc, Y. Lipid peroxidation and antioxidative enzyme activities in childhood epilepsy. *J. Child Neurol.* **2002**, *17*, 673–676. [CrossRef] [PubMed]
37. Verrotti, A.; Basciani, F.; Trotta, D.; Pomilio, M.P.; Morgese, G.; Chiarelli, F. Serum copper, zinc, selenium, glutathione peroxidase and superoxide dismutase levels in epileptic children before and after 1 year of sodium valproate and carbamazepine therapy. *Epilepsy Res.* **2002**, *48*, 71–75. [CrossRef]
38. Gunes, S.; Dirik, E.; Yis, U.; Seckin, E.; Kuralay, F.; Kose, S.; Unalp, A. Oxidant status in children after febrile seizures. *Pediatr. Neurol.* **2009**, *40*, 47–49. [CrossRef] [PubMed]
39. Biswas, S.K.; de Faria, J.B. Which comes first: Renal inflammation or oxidative stress in spontaneously hypertensive rats? *Free Radic. Res.* **2007**, *41*, 216–224. [CrossRef] [PubMed]
40. Fischer, R.; Maier, O. Interrelation of oxidative stress and inflammation in neurodegenerative disease: Role of tnf. *Oxid. Med. Cell. Longev.* **2015**, *2015*, 610813. [CrossRef] [PubMed]
41. Pashkow, F.J. Oxidative stress and inflammation in heart disease: Do antioxidants have a role in treatment and/or prevention? *Int. J. Inflamm.* **2011**, *2011*, 514623. [CrossRef] [PubMed]
42. Petersen, K.S.; Smith, C. Ageing-associated oxidative stress and inflammation are alleviated by products from grapes. *Oxid. Med. Cell. Longev.* **2016**, *2016*, 6236309. [CrossRef] [PubMed]
43. Lassmann, H.; van Horssen, J. Oxidative stress and its impact on neurons and glia in multiple sclerosis lesions. *Biochim. Biophys. Acta* **2016**, *1862*, 506–510. [CrossRef] [PubMed]
44. Biswas, S.K. Does the interdependence between oxidative stress and inflammation explain the antioxidant paradox? *Oxid. Med. Cell. Longev.* **2016**, *2016*, 5698931. [CrossRef] [PubMed]
45. Flohe, L.; Brigelius-Flohe, R.; Saliou, C.; Traber, M.G.; Packer, L. Redox regulation of nf-kappa b activation. *Free Radic. Biol. Med.* **1997**, *22*, 1115–1126. [CrossRef]
46. Lorigados Pedre, L.; Morales Chacon, L.M.; Pavon Fuentes, N.; Robinson Agramonte, M.L.A.; Serrano Sanchez, T.; Cruz-Xenes, R.M.; Diaz Hung, M.L.; Estupinan Diaz, B.; Baez Martin, M.M.; Orozco Suárez, S. Follow-up of peripheral il-1beta and il-6 and relation with apoptotic death in drug-resistant temporal lobe epilepsy patients submitted to surgery. *Behav. Sci. (Basel, Switzerland)* **2018**, *8*, 21.
47. Keskin Guler, S.; Aytac, B.; Durak, Z.E.; Gokce Cokal, B.; Gunes, N.; Durak, I.; Yoldas, T. Antioxidative-oxidative balance in epilepsy patients on antiepileptic therapy: A prospective case-control study. *Neurol. Sci.* **2016**, *37*, 763–767. [CrossRef] [PubMed]
48. Donmezdil, N.; Cevik, M.U.; Ozdemir, H.H.; Tasin, M. Investigation of pon1 activity and mda levels in patients with epilepsy not receiving antiepileptic treatment. *Neuropsychiatr. Dis. Treat.* **2016**, *12*, 1013–1017. [PubMed]
49. Yehuda, S.; Rabinovitz, S.; Carasso, R.L.; Mostofsky, D.I. The role of polyunsaturated fatty acids in restoring the aging neuronal membrane. *Neurobiol. Aging* **2002**, *23*, 843–853. [CrossRef]
50. Farooqui, A.A.; Horrocks, L.A. Lipid peroxides in the free radical pathophysiology of brain diseases. *Cell. Mol. Neurobiol.* **1998**, *18*, 599–608. [CrossRef] [PubMed]
51. Leonarduzzi, G.; Arkan, M.C.; Basaga, H.; Chiarpotto, E.; Sevanian, A.; Poli, G. Lipid oxidation products in cell signaling. *Free Radic. Biol. Med.* **2000**, *28*, 1370–1378. [CrossRef]

52. Poli, G.; Biasi, F.; Leonarduzzi, G. 4-hydroxynonenal-protein adducts: A reliable biomarker of lipid oxidation in liver diseases. *Mol. Asp. Med.* **2008**, *29*, 67–71. [CrossRef] [PubMed]

53. Zarkovic, N.; Zarkovic, K.; Kralj, M.; Borovic, S.; Sabolovic, S.; Blazi, M.P.; Cipak, A.; Pavelic, K. Anticancer and antioxidative effects of micronized zeolite clinoptilolite. *Anticancer Res.* **2003**, *23*, 1589–1595. [PubMed]

54. Pecorelli, A.; Ciccoli, L.; Signorini, C.; Leoncini, S.; Giardini, A.; D'Esposito, M.; Filosa, S.; Hayek, J.; De Felice, C.; Valacchi, G. Increased levels of 4hne-protein plasma adducts in rett syndrome. *Clin. Biochem.* **2011**, *44*, 368–371. [CrossRef] [PubMed]

55. Pecorelli, A.; Leoncini, S.; De Felice, C.; Signorini, C.; Cerrone, C.; Valacchi, G.; Ciccoli, L.; Hayek, J. Non-protein-bound iron and 4-hydroxynonenal protein adducts in classic autism. *Brain Dev.* **2013**, *35*, 146–154. [CrossRef] [PubMed]

56. Frantseva, M.V.; Perez Velazquez, J.L.; Tsoraklidis, G.; Mendonca, A.J.; Adamchik, Y.; Mills, L.R.; Carlen, P.L.; Burnham, M.W. Oxidative stress is involved in seizure-induced neurodegeneration in the kindling model of epilepsy. *Neuroscience* **2000**, *97*, 431–435. [CrossRef]

57. Dalleau, S.; Baradat, M.; Gueraud, F.; Huc, L. Cell death and diseases related to oxidative stress: 4-hydroxynonenal (hne) in the balance. *Cell Death Differ.* **2013**, *20*, 1615–1630. [CrossRef] [PubMed]

58. Mark, R.J.; Lovell, M.A.; Markesbery, W.R.; Uchida, K.; Mattson, M.P. A role for 4-hydroxynonenal, an aldehydic product of lipid peroxidation, in disruption of ion homeostasis and neuronal death induced by amyloid beta-peptide. *J. Neurochem.* **1997**, *68*, 255–264. [CrossRef] [PubMed]

59. Montine, K.S.; Reich, E.; Neely, M.D.; Sidell, K.R.; Olson, S.J.; Markesbery, W.R.; Montine, T.J. Distribution of reducible 4-hydroxynonenal adduct immunoreactivity in alzheimer disease is associated with apoe genotype. *J. Neuropathol. Exp. Neurol.* **1998**, *57*, 415–425. [CrossRef] [PubMed]

60. Theard, M.A.; Baughman, V.L.; Wang, Q.; Pelligrino, D.A.; Albrecht, R.F. The role of nitric oxide in modulating brain activity and blood flow during seizure. *Neuroreport* **1995**, *6*, 921–924. [CrossRef] [PubMed]

61. Bosnak, M.; Ayyildiz, M.; Yildirim, M.; Agar, E. The role of nitric oxide in the anticonvulsant effects of pyridoxine on penicillin-induced epileptiform activity in rats. *Epilepsy Res.* **2007**, *76*, 49–59. [CrossRef] [PubMed]

62. Sardo, P.; Ferraro, G. Modulatory effects of nitric oxide-active drugs on the anticonvulsant activity of lamotrigine in an experimental model of partial complex epilepsy in the rat. *BMC Neurosci.* **2007**, *8*, 47. [CrossRef] [PubMed]

63. Osonoe, K.; Mori, N.; Suzuki, K.; Osonoe, M. Antiepileptic effects of inhibitors of nitric oxide synthase examined in pentylenetetrazol-induced seizures in rats. *Brain Res.* **1994**, *663*, 338–340. [CrossRef]

64. Murashima, Y.L.; Yoshii, M.; Suzuki, J. Ictogenesis and epileptogenesis in el mice. *Epilepsia* **2002**, *43* (Suppl. 5), 130–135. [CrossRef] [PubMed]

65. Sawa, T.; Akaike, T.; Maeda, H. Tyrosine nitration by peroxynitrite formed from nitric oxide and superoxide generated by xanthine oxidase. *J. Biol. Chem.* **2000**, *275*, 32467–32474. [CrossRef] [PubMed]

66. Ryan, K.; Liang, L.P.; Rivard, C.; Patel, M. Temporal and spatial increase of reactive nitrogen species in the kainate model of temporal lobe epilepsy. *Neurobiol. Dis.* **2014**, *64*, 8–15. [CrossRef] [PubMed]

67. Solowiej, E.; Sobaniec, W. The effect of antiepileptic drug therapy on antioxidant enzyme activity and serum lipid peroxidation in young patients with epilepsy. *Neurol. Neurochir. Pol.* **2003**, *37*, 991–1003. [PubMed]

68. Peker, E.; Oktar, S.; Ari, M.; Kozan, R.; Dogan, M.; Cagan, E.; Sogut, S. Nitric oxide, lipid peroxidation, and antioxidant enzyme levels in epileptic children using valproic acid. *Brain Res.* **2009**, *1297*, 194–197. [CrossRef] [PubMed]

69. Martinc, B.; Grabnar, I.; Vovk, T. The role of reactive species in epileptogenesis and influence of antiepileptic drug therapy on oxidative stress. *Curr. Neuropharmacol.* **2012**, *10*, 328–343. [CrossRef] [PubMed]

70. Banach, M.; Piskorska, B.; Czuczwar, S.J.; Borowicz, K.K. Nitric oxide, epileptic seizures, and action of antiepileptic drugs. *CNS Neurol. Disord. Drug Targets* **2011**, *10*, 808–819. [CrossRef] [PubMed]

71. Menon, B.; Ramalingam, K.; Kumar, R.V. Oxidative stress in patients with epilepsy is independent of antiepileptic drugs. *Seizure* **2012**, *21*, 780–784. [CrossRef] [PubMed]

behavioral sciences

MDPI

Article

Follow-Up of Peripheral IL-1β and IL-6 and Relation with Apoptotic Death in Drug-Resistant Temporal Lobe Epilepsy Patients Submitted to Surgery

Lourdes Lorigados Pedre [1,*]**, Lilia M. Morales Chacón** [2]**, Nancy Pavón Fuentes** [1]**,**
María de los A. Robinson Agramonte [1]**, Teresa Serrano Sánchez** [1]**, Rachel M. Cruz-Xenes** [3]**,**
Mei-Li Díaz Hung [1]**, Bárbara Estupiñán Díaz** [4]**, Margarita M. Báez Martín** [2] **and**
Sandra Orozco-Suárez [5]

[1]   Immunochemical Department, International Center for Neurological Restoration, 25th Ave, Playa, 15805,
       PC 11300 Havana, Cuba; nancy@neuro.ciren.cu (N.P.F.); robin@neuro.ciren.cu (M.d.l.A.R.A.);
       teresa@neuro.ciren.cu (T.S.S.); mdiazhung@gmail.com (M.L.D.H)
[2]   Clinical Neurophysiology Lab., International Center for Neurological Restoration, PC 11300 Havana, Cuba;
       lily@neuro.ciren.cu (L.M.M.C.); minou@neuro.ciren.cu (M.M.B.M.)
[3]   Biology Faculty, University of Havana, PC 10400 Havana, Cuba; rachel.cruz@fbio.uh.cu
[4]   Morphological Laboratory, International Center for Neurological Restoration, PC 11300 Havana, Cuba;
       baby@neuro.ciren.cu
[5]   Unit of Medical Research in Neurological Diseases, Specialty Hospital, National Medical Center,
       XXI Century IMSS, PC 06720 Mexico City, Mexico; sorozco5@hotmail.com
*   Correspondence: lourdes.lorigados@infomed.sld.cu; Tel.: +53-7-2715353

Received: 5 December 2017; Accepted: 30 January 2018; Published: 5 February 2018

Abstract: Increasing amounts of evidence support the role of inflammation in epilepsy. This study was done to evaluate serum follow-up of IL-1β and IL-6 levels, as well as their concentration in the neocortex, and the relationship of central inflammation with NF-κB and annexin V in drug-resistant temporal lobe epileptic (DRTLE) patients submitted to surgical treatment. Peripheral and central levels of IL-1β and IL-6were measured by ELISA in 10 DRTLE patients. The sera from patients were taken before surgery, and 12 and 24 months after surgical treatment. The neocortical expression of NF-κB was evaluated by western blotting and annexin V co-localization with synaptophysin by immunohistochemistry. The neocortical tissues from five patients who died by non-neurological causes were used as control. Decreased serum levels of IL-1 and IL-6 were observed after surgery; at this time, 70% of patients were seizure-free. No values of IL-1 and IL-6 were detected in neocortical control tissue, whereas cytokine levels were evidenced in DRTLE. Increased NF-κB neocortex expression was found and the positive annexin V neurons were more obvious in the DRTLE tissue, correlating with IL-6 levels. The follow-up study confirmed that the inflammatory alterations disappeared one year after surgery, when the majority of patients were seizure-free, and the apoptotic death process correlated with inflammation.

Keywords: drug-resistant temporal lobe epilepsy; inflammation; apoptosis; IL-1β; IL-6; NF-κB

1. Introduction

Epilepsy is a chronic brain disease that affects around 50 million people worldwide. In a high number of cases it is characterized by the presence of recurrent seizures, which are the result of an excessive electrical discharge of a neuronal group in a certain part of the brain [1]. The battery of treatments designed to counteract the clinical manifestations of this disease is diverse and ranges from a wide spectrum of antiepileptic drugs, specific diets, and sports-based therapies, to surgical techniques for resection of the epileptogenic focus [2]. Despite many efforts to find a successful pharmacological treatment for this condition, 30% of the population of patients with epilepsy cannot control the onset of seizures [3]. This clinical condition is known as drug-resistant epilepsy (DRE) or intractable epilepsy [4]. The literature has reported that temporal lobe epilepsy (TLE) shows one of the highest incidences of DRE [5]. Surgical treatment is currently an option for patients with drug-resistant temporal lobe epilepsy (DRTLE) and is the only treatment that clearly shows a positive action in decreasing the frequency of seizures and the progression of the disease in these patients [6].

Several authors suggest the contribution of immunological and inflammatory mechanisms in the pathogenesis of epilepsy based on the favorable effects of treatment with intravenous immunoglobulins, corticosteroids, and anti-inflammatories [7–10]. The immunological alterations described in epilepsy are associated, in most cases, with anti-epileptic drug treatment [11,12]; in others, they have not been related to pharmacotherapy [13,14]. Additionally, our group has described that immunological disorders in patients with epilepsy are associated with certain locations of the epileptogenic zone. Temporal lobe localization of the epileptogenic zone has been shown to be related to alterations in cellular immunity, and this dysfunction is not associated with pharmacological antiepileptic treatment [15,16]. However, currently, it has not been clarified whether inflammatory and immunological disorders are cause or consequence of the seizures in DRTLE.

Inflammation is considered an important factor in the pathophysiology of seizures. All the risk factors for epilepsy such as traumas, tumors, and infections are accompanied by different degrees of inflammation in the central nervous system, which is associated with the occurrence of seizures [17–19]. However, little is known about the contribution of peripheral inflammation on the modulation of these events as well as on the relationship of the neuronal loss described in epilepsy with inflammatory processes.

Proinflammatory cytokines are highly studied markers in both patients and experimental models of drug-resistant epilepsy. IL-1 and IL-6 are two of the most commonly approached proinflammatory cytokines in studies conducted to evaluate inflammation in drug-resistant epilepsy [20–26]. However, as far as we know, follow-up studies of the concentrations of inflammatory markers have not been described for a period of up to 2 years after removal of the epileptogenic focus by surgical techniques (standard lobectomies guided by electrocorticography).

The main objective of this work is to evaluate the proinflammatory proteins after the elimination of the epileptogenic focus and its relation with the neuronal loss described in patients with DRTLE.

2. Materials and Methods

2.1. Patient Information

Ten patients from the International Center for Neurological Restoration (CIREN) with DRTLE (6 female and 4 male, with mean age 33.1 ± 6.35 years) who underwent epilepsy surgery were included for the study of serum concentration of inflammatory cytokines (IL-1β and IL-6). The criteria to be considered drug-resistant were the following: present seizures for more than two years, two complex partial monthly seizures, use of two first-line anti-epileptic drugs, two cycles of monotherapy and at least one of polytherapy.

All patients were evaluated in the Video-EEG Telemetry Unit from CIREN, and they underwent a complete general and neurological physical examination and anatomical evaluation by Magnetic Resonance Imaging (1.5 T MAGNETOM SINPHONY) and SPECT.

Temporal localization presents unilateral rhythmic unloading with maximum amplitude in the zygomatic electrodes and in the anterior or middle temporal ones as the first electrographic change.

Antiepileptic treatment: The most commonly used drug was Carbamazepine (3–6 tablets daily) followed by Clobazan (3–4 tablets daily), Magnesium Valproate (1–6 tablets daily), and Lamotrigine (2–3 tablets daily). Up to two years after the surgery, all patients received the same drug treatment.

The neocortex adjacent to the hippocampus of the patients was resected at the time of surgery and the resection of the epileptic zone was performed by means of standard temporal lobectomies adjusted by electrocorticography.

The neocortical tissues from patients in this study were evaluated to determine the concentrations of IL-1β and IL-6, co-localization of annexinV and synaptophysine, and the expression of NF-κB. The neocortex control tissues were obtained from subjects who died due to non-neurological causes (Department of Pathological Anatomy from Clinical Hospital, Havana, Cuba). The control tissues were matched in age and gender with the group of DRTLE patients (average age: 35.2 years, 3 male and 2 female). They did not show macro- or microscopic pathological abnormalities, and the mean time of obtaining tissue was 3.2 h after the subject died.

2.2. Samples

The collected clinical parameters included demographic characteristics and clinical state showed in Table 1. The parameters were age; gender; personal pathological history; side of focus; disease duration; and the clinical state of patients, according to Engel's scale, one and two years after surgery. Post-surgical seizure outcome assessment was based on the system proposed by Engel. (Engel class I, free of disabling seizures; class IA, seizure-free; class II, rare seizures (fewer than three seizures per year); class III, worthwhile improvement (reduction in seizures of 80% or more); class IV, no benefit [27].)

Table 1. Summary of clinical data from control subjects and postmortem interval.

Control	Cause of Death	Postmortem Interval (h)
1	Pulmonary thromboembolism	2.5
2	Vehicle accident	3
3	Vehicle accident	4
4	Vehicle accident	3
5	Vehicle accident	3.5

Blood: Venous Blood samples (5 mL) of the patients were taken by antecubital puncture, after asepsis of the region. The blood was centrifuged and the serum was kept at −20 °C until use. The samples of serum were obtained before and after surgical treatment (1 and 2 years of surgical evolution time).

Neocortical tissue: The neocortical tissue from patients with DRTLE was obtained during surgery. The tissue was washed with 0.9% saline solution at 4 °C, a fragment of approximately 0.5 cm^3 was cut, placed in liquid nitrogen, and later transferred to a −80 °C freezer where it was preserved until use.

The neocortical control tissue was obtained from necropsy and was subsequently fractionated by areas. A fragment of approximately 0.5 cm^3 was taken from the neocortex; this fragment was processed in a similar way to patient's tissues.

Control samples were obtained from five subjects who died by vehicle accident ($n = 4$) and pulmonary thromboembolism ($n = 1$), and without history of neurological disease (Table 1). The neocortical tissue was dissected at the time of autopsy with a postmortem interval (PMI) from 2.5 to 4 h after death, and the samples were immediately stored at −70 °C. Previous works that used autopsy tissues with a PMI longer than the samples of the present work have shown the preservation of the protein and mRNA [28–30].

Homogenization of brain tissue: Tissue samples were homogenized in a glass–Teflon (Potter-Elvehjem), containing lysis buffer as described by the manufacturer (ENZO Life Sciences, Llorach, Germany)for the cytokine ELISA and 1mL of Trizol reagent (Invitrogen, Carlsbad, CA, USA) for the western blot. Protein was isolated following the manufacturer's instructions.

2.3. Immunoenzymatic Assay for IL-1β and IL-6 in Serum and Brain Tissue

The concentrations of IL-1β and IL-6 (ENZO Life Sciences, Llorach, Germany) were measured by ELISA according to manufacturer's instructions. The concentrations of IL-1β and IL-6 from the homogenate of the neocortical tissue of patients with DRTLE and control subjects was measured by the same methods. The lower limit of detection was 6pg/mL for IL-6 and 1 pg/mL for IL-β. Briefly, serum or homogenate from neocortical tissue was incubated in coated 96-well plates at room temperature for 2 h. The serum and tissue samples were applied in duplicate as well as the dilutions from the standard curves (IL-1β and IL-6). Plates were washed and then incubated with the detection antibody. After rewashing the plates, the conjugate was added for 30 min, followed by the substrate solution. The reaction was stopped with 1N H_2SO_4 and optical densities were measured at 450 nm using a microplate reader (ELx800 BioTek Instruments, Inc., Winooski, VT, USA).

2.4. Extraction of Neocortical Tissue Proteins and Western Blot to Evaluate NF-κB

The expression levels of NF-κB in the homogenates were analyzed by gel electrophoresis and western blot. The concentration of total proteins was determined by Lowry's method [31]. The absorbance values at 630 nm were obtained by spectrophotometry (BioTek Instruments, Winooski, VT, USA). Subsequently, the samples were separated by electrophoretic methods in SDS polyacrylamide gel. The proteins were transferred by electrophoresis to vinylidene membranes (Sigma-Aldrich, St. Louis, MO, USA). The membranes were blocked with 2% milk in phosphate buffer solution/Triton X-100 to avoid nonspecific binding. After blocking, the membranes were incubated individually with anti-NF-κB antibody and rabbit polyclonal anti-GAPDH (Glyceraldehyde-3-Phosphate Dehydrogenase, Santa Cruz Biotechnology, Santa Cruz, CA, USA) overnight with mild shaking at 4 °C (BrinkmannOrbMix 110, Brinkmann, Germany). The membranes were incubated for 2 h at room temperature in phosphate buffer containing goat anti-mouse IgG/rabbit conjugated with horseradish peroxidase diluted 1:1000 (Cell Signaling, Oregon City, OR, USA) for 1 h. The immunoreactive bands were detected by improved chemiluminescence. The densities of the protein signals in the films were quantified using ImageJ software.

To determine the integrity of the extracted proteins, 1D SDS-PAGE electrophoresis (12%) was used, followed by staining with Coomassie blue and western blot to identify the constitutive protein, indicating that there is no protein degradation.

2.5. Immunohistochemistry to Annexin V and Synaptophysin

The fragment of neocortical tissue obtained during surgery was washed with 0.9% saline at 4 °C and fixed in phosphate solution containing 4% paraformaldehyde and 0.2% glutaraldehyde. Subsequently, they were submerged in increasing solutions of sucrose (15, 20, 25, and 30%) and frozen at −70 °C. The neocortex tissue was cut into sections of 10 μm (Cryostat Leitz, 1720, Wetzlar, Germany). The pieces of tissue were blocked and incubated with the apoptotic marker annexin V and neuronal marker synaptophysin (Santa Cruz Biotechnology, Santa Cruz, CA, USA) according to the instructions of the manufacturer. An antibody conjugated with FITC (Zymed Laboratories Inc., San Francisco, CA, USA) was used for the immunodetection of antibodies against annexin V, while the immunodetection of synaptophysin was carried out by means of an antibody conjugated with Alexa Fluor 647 (Lab. Mol. Probes, Oregon City, OR, USA). The co-localization of the apoptotic and neuronal markers was evaluated in each patient's tissue. The nuclei were visualized by counterstaining with propidium iodide. All sections were examined in a confocal microscope (Bio-Rad, Cambridge, UK) and each plate was examined by two specialists (SO and LL).

Quantitative Analysis: Only the sections that had consistent immunostaining were counted and 5 to 10 sections of each patient or control were quantified. The percentage of immunopositive cells to each marker (number of immunoreactive cells per mm^3) was calculated in relation to the number of cells stained with propidium iodide per mm³, and the result was expressed as the percentage of cells positive for each marker. All sheets were examined by two specialists (SO and LL).

2.6. Ethical Considerations

All procedures followed the rules of the Declaration of Helsinki of 1975 for human research, and the study was approved by Record 03/2015, given by the Ethics Committee of the International Center for Neurological Restoration, named according to RESOLUTION No. 29-P/2011, on 24 October 2011. Each patient or family gave their informed consent.

2.7. Statistical Processing

Statistical analysis was carried out using the GraphPad Prism 5 software (GraphPad Software, Inc., La Jolla, CA, USA). The values are expressed as mean $\pm$ SEM. Normal distribution and homogeneity of variance of the data were tested by the Kolmogorov–Smirnov and Levene tests, respectively. The comparisons between two groups were made by means of the *t*-Student test while comparisons between more than two groups were made by one-way ANOVA, with post hoc Dunnet test. The Pearson correlation was used for the correlation study. In all cases, statistically significant differences were considered when $p \leq 0.05$.

3. Results

3.1. Effect of Surgical Treatment in DRTLE Patients on the Peripheral (Serum) and Central (Neocortical Tissue) Concentrationsof IL-1βand IL-6.

In order to evaluate the effect on serum concentrations of IL-1β and IL-6 in DRTLE patients who received surgical treatment, we measured the concentrations of these cytokines before, and one and two years after the resection of the epileptogenic zone. In this study, there are no gender differences inboth levels of interleukins (IL-1 and IL-6), so the concentrations in male and female sexes were unified in one group. The results showed a statistically significant decrease of both (IL-1β and IL-6) cytokines (Figure 1A,B)one and two years after surgery evolution. In the first post-surgical evaluation, 70% of patients were in category IA according to the Engel scale (seizure-free), and those remaining had a significant decrease in seizure frequency (Table 2). One year after surgery treatment, lower values of IL-1β and IL-6 were observed in seizure-free patients, while patients who remained with seizures showed higher values of both cytokines (Figure 2A,B).

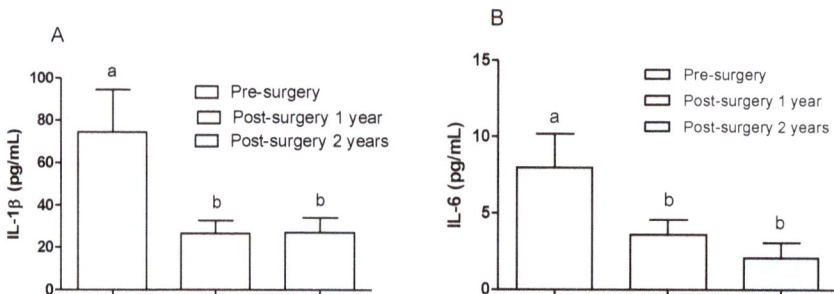

Figure 1. Comparison of IL-1β and IL-6 concentrations before and after surgical treatment. (**A**) IL-1β serum concentration. (**B**) IL-6 serum concentration. The bars represent the mean $\pm$ the standard error of the mean. One way ANOVA, post hoc Dunnet test, $p \leq 0.05$.

Table 2. Clinical data from patients with drug-resistant temporal lobe epilepsy submitted to temporal lobectomy.

Patient Number	Age (Year)	Gender	Personal Pathological History	Side of Focus (R, L)	Disease Duration	Pre-Surgery Number of Seizures per Month	Engel Scale	
							One Year Post-Surgery	Two Years Post-Surgery
1	23	F	Bronchial asthma	L	16	11	IIIA	IIIA
2	41	F	No history	L	15	15	IIIA	IIIA
3	35	F	Febrile seizures	R	35	12	IA	IA
4	26	F	Cranioencephalic trauma	R	11	13	IA	IA
5	35	M	Meningoencephalitis	R	34	12	IA	IA
6	31	M	Febrile seizures	R	30	10	IA	IA
7	41	M	Bronchial asthma	L	36	21	IA	IA
8	26	F	Febrile seizures	L	26	15	IIA	IIIA
9	36	M	Cranioencephalic trauma	R	32	12	IA	IA
10	37	F	Bronchial asthma	R	21	11	IA	IA

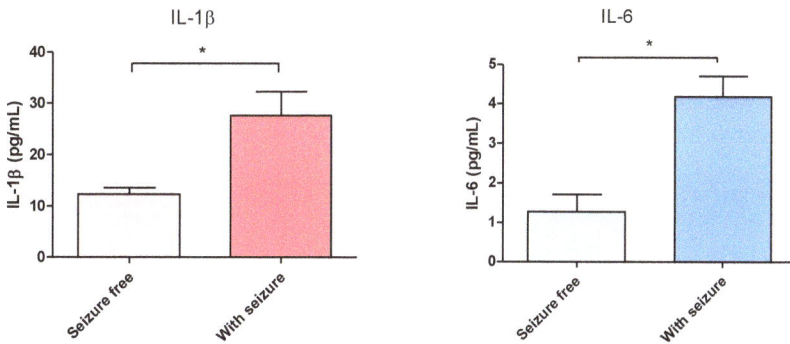

Figure 2. Concentration in serum of Interleukins according to the presence or absence of seizures one year after surgery. (**A**) IL-1β. (**B**) IL-6. The bars represent the mean ± the standard error of the mean. Mann Whitney test, * $p \leq 0.048$.

Neocortical tissue from DRTLE patients showed values of IL-1β between 9.69 and 78.32 pg/mL with a mean value of 21.58 pg/mL (Figure 3A), and IL-6 between 17.53 and 278.3 pg/mL with a mean value of 78.87 pg/mL (Figure 3B). The concentrations of both cytokines in the control group presented values below the limit of detection for the method used (Figure 3A,B).

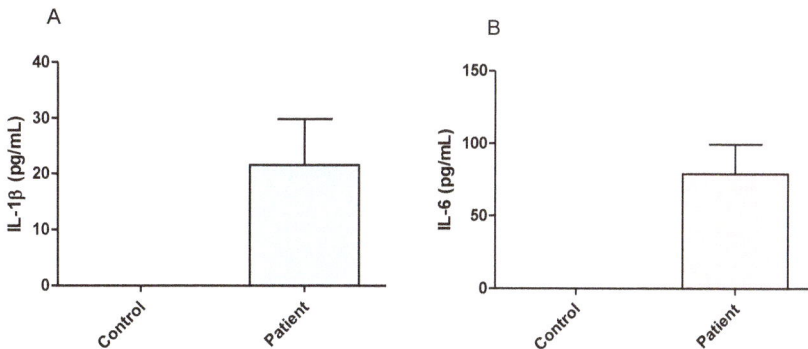

Figure 3. Concentrations of IL-1β and IL-6 in neocortical tissue from patients with drug-resistant temporal lobe epilepsy and control tissue. (**A**) Concentrations of IL-1β. (**B**) Concentrations of IL-6. The bars represent the mean ± the standard error of the mean.

3.2. Neocortical Tissue Expression of NF-κB

The presence of the molecule involved in the inflammatory cascade and linked to death processes in neocortical tissue, NF-κB, was evaluated, and the results are shown in Figure 4. Figure 4A shows a statistically significant increase in the expression of NF-κB in the group of patients with DRTLE compared with in the control subjects, while Figure 4B represents an example of the western blot bands in a patient with DRTLE and a control subject, as well as those for GAPDH.

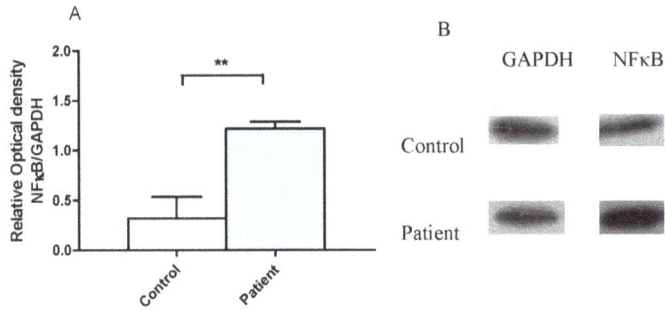

Figure 4. Expression of NF-κB in neocortical tissue of patients with drug-resistant temporal lobe epilepsy. (**A**) Comparison of relative optical density values of NF-κB in patients and control subjects. (**B**) Representative example of the immunodetection of NF-κB in a patient and a control, as well as for GAPDH. The bars represent the mean $\pm$ the standard error of the mean, *t*-Student test, ** $p \leq 0.001$

3.3. Relationship between Central Inflammation and Apoptotic Neuronal Death

Figure 5 shows an increase in co-localization of annexin V and synaptophysin positive cells in DRTLE patients in comparison with in control tissue. It is important to note that there is a lower number of positive synaptophysin cells in patients in relation to in the control subjects.

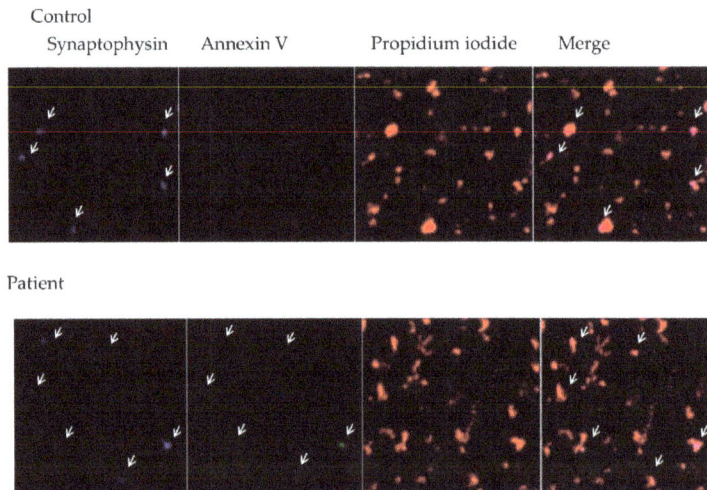

Figure 5. Confocal image of the neocortex of a control subject and a patient illustrating the immunodetection of annexin V + cells (green) doubly marked with synaptophysin (blue) and counterstained with propidium iodide (red). Note that the patient's annexin V + cells coincide with synaptophysin + (40×).

There is a positive correlation between the percentage of immunopositive cells to annexin V and the concentration of IL-6 in neocortical tissue (Figure 6).

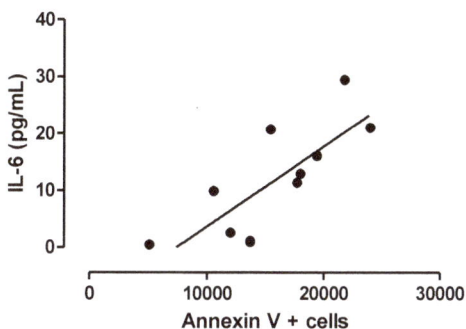

Figure 6. Correlation between the number of annexin V positive cells and IL-6 concentration. Spearman, $r = 0.8519$, $p \leq 0.001$.

4. Discussion

4.1. Inflammatory Cytokines (IL-1β and IL-6)

The concept of "immunological privilege" of the CNS has been eliminated because of, among other reasons, the accumulation of evidence that immune responses and inflammatory reactions are involved in the pathogenesis of several brain diseases like epilepsy [32]. There is evidence of cerebral inflammation associated with drug-resistant epilepsy of various aetiologies in patients who underwent surgery and whose brain tissues had proinflammatory molecules, reactive astrocytosis, activated microglia, and other indicators of inflammation in the hippocampus [33]. There are several experimental studies that show that the activity of epileptic seizures by themselves can induce cerebral inflammation and that the recurrence of these seizures prolongs chronic inflammation. Similarly, the action of inflammatory mediators on the generation of seizures has been proposed [34,35].

The study of different models of epilepsy developed to evaluate the inflammatory response has shown that, in rodents, with convulsive activity occurs the following molecular cascade: a rapid increase in the proinflammatory cytokines (IL-1β, IL-6, and TNF-α), the activation of the signaling pathway of Toll-like receptors (TLRs), the production of chemokines, activation of the complement system, and increased expression of adhesion molecules [23,25,36–39].

On the other hand, the literature supports that the immunological and inflammatory alterations described in epilepsy may be associated with anti-epileptic drugs. Changes in inflammatory markers related to carbamazepine, valproate, and other drugs are described [40,41].

Serum levels of IL-1β and IL-6 before and after surgery were evaluated in order to detect if there were changes in the peripheral proinflammatory cytokines from DRTLE patients and if these changes were modified once the epileptogenic zone was resected. This follow-up study was related to the clinical evolution of these patients in terms of the presence or absence of seizures one and two years after surgery. The decrease in the concentration of IL-1β and IL-6 one year after surgery and the fact that in this period most of the patients are free of seizures supports the idea that seizures are the cause of the inflammatory disorders observed in patients with DRTLE. Interestingly, 70% of patients remained seizure-free one year after surgery and the remainder (25%) showed a substantial reduction in seizure frequency.

According to our results, the low levels of IL-6 in patients who were free of seizures one year after surgery in comparison with those in the patients who still experienced them allow us to confirm that this inflammatory marker is closely linked with the occurrence of seizures. Our results indicate that once the epileptogenic zone is resected and seizure activity is reduced, there is a decrease in proinflammatory cytokines. These findings may suggest that seizures are the cause of inflammatory disorders observed in patients with DRTLE.

Other authors report an increase of IL-6 concentrations in patients with epilepsy compared with those in control subjects. They describe this increase associated with the temporal location of the epileptogenic zone [42]. Coincidentally, previous studies by our group showed that alterations in cellular immunity are associated with the location of the epileptogenic zone. The immune alteration was observed in temporal but not in extratemporal localization [43].

An important aspect to assess is the integration of inflammatory events that occur both peripherally and centrally. The existence of several mechanisms by which peripheral inflammation interacts with the brain and induces inflammation at the central level has been discussed. The immediate recognition of inflammatory mediators such as cytokines can occur in circumventricular organs due to the expression of TLRs and IL-R in these structures [44]. The cytokine signal is another mechanism that mediates the response in the brain to peripheral inflammation. Cytokines use both the neural pathway (vagus afferents) and the humoral pathway to communicate the innate immune system to the brain. These mediators can also be transported through the blood–brain barrier (BBB) [45–47], and these cytokine signals trigger a central inflammatory response with activation of the microglia, which that induce and propagate these inflammatory signals in the CNS [48].

The evaluation of IL-1β and IL-6 levels in brain tissue (neocortex) of patients with DRTLE showed values that indicate the presence of an inflammatory process at the central level reinforced by the absence of these markers in the control tissue. Tissue obtained at autopsy from subjects with no evidence of neurological disease was used as controls since previous reports indicate that proteins from brain tissue are preserved for several hours after death [28–30,49]. Previously, other authors supported the nondetection of concentrations of proinflammatory cytokines in the brain tissue of control subjects [50]. Coincidentally, in resected tissue of patients with TLE, other investigators have suggested that the inflammatory pathways are activated during epileptogenesis and that they persist in epileptic tissue, which contributes to the etiopathogenesis of TLE [34,51].It is also known that inflammatory activity is affected in different ways depending on the severity of the seizures [52,53].

In particular, the analysis of the mRNA of IL-1β and IL-1Ra after the systemic injection of kainic acid in rats shows a significant induction in the microglial cells of the hippocampus, as well as in other areas of the limbic system [54]. Vezzani et al. provided evidence of the rapid increase in IL-1β levels in the hippocampus after seizures in the model with kainic acid; they assume that it occurs as a result of the activation of microglial cells [55]. The increase of IL-1β, IL-6, and TNF-α in the microglia and astrocytes is followed by a cascade of inflammatory events that can recruit cells of the adaptive immune system [33].

Indistinctly in blood or cerebrospinal fluid, an increase in IL-1β and its receptor in astrocytes, microglia, and neurons was documented in a series of patients surgically treated for TLE with hippocampal sclerosis. IL-1β and its receptor are highly expressed in the neurons and glia of patients with TLE caused by focal cortical dysplasia and neuroglial tumors, where expression in brain tissue without these alterations is negligible [34].

In summary, we can affirm that there are high IL-1βand IL-6 concentrations at both the systemic and the central levels that sustain the presence of an inflammatory response in DRTLE. The decrease of both cytokines after surgery in patients with a drastic reduction in seizures supports the idea that these events are an important cause of inflammation.

4.2. Expression of Molecules Associated withInflammation (NF-κB) and Apoptotic Death in Brain Tissue from DRTLE Patients

Previous studies report neuronal death in the neocortical tissue of patients with DRTLE [56,57], making it necessary to evaluate the expression of a molecule that participates in the death mechanism by apoptosis and which is linked to inflammatory processes: NF-κB.

Increased expression of NF-κB in the neocortical tissue of patients with DRTLE was observed. This fact confirms the activation of the inflammatory response at the central level and helps to explain the presence of apoptotic neuronal death observed in this study.

NF-κB is a key mediator of the innate and inflammatory immune response; it is activated by inflammatory cytokines through the TNF receptor, IL-1 receptor, and the TLRs family, and, in turn, their activation gives rise to the expression of other cytokines, proteases, and metabolic enzymes [58].

NF-κB has been shown to be involved in the neuropathological processes associated with seizures in epilepsy. Epileptic seizures have been shown to trigger a rapid increase in the amount of NF-κB activated in the hippocampus [59] and its over-expression has been demonstrated in reactive astrocytes and surviving neurons of patients with hippocampal sclerosis [60].

There are contradictory criteria related to the existence or otherwise of other affected areas different to the mesial regions in DRTLE [61]. Previous results from our work group support neuronal death in layer IV of the neocortex adjacent to the hippocampus. This layer is made up of gabaergic interneurons, neurons that are especially sensitive to damage, and glutamatergic-type afferents that explain the occurrence of death processes due to excitotoxicity [62].

Annexin V is described as an early marker of apoptotic processes. Once the process of apoptosis in a cell is activated, the externalization of phosphatidyl serine that is distributed asymmetrically in the cell membrane is produced and is able to bind annexin V. In our study, increased immunoreaction to coincidental annexin V was detected with immunodetection to synaptophysin in the neocortex of patients with DRTLE, which confirms the presence of neurons that initiate the apoptotic process in this tissue.

The neuronal death observed in our results is probably due to the inflammatory process and to the involvement or mitochondrial dysfunction caused by the depolarization of the membrane and the loss of permeability of the mitochondrial membrane, a fact that usually implies the death of the cell [63]. Prolonged exposure to reactive oxygen species can cause depolarization of the mitochondrial membrane and subsequently result in an alteration of the permeability of this structure. Our working group has previously described the presence of an imbalance of the redox system in this type of patient [64]. Another possibility could be the release by the mitochondria of the inhibiting factor of apoptosis that can directly cause chromosomal damage and/or the reactive oxygen species of lysosomal cathepsins that can also provoke mitochondrial damage [63].

The relevant role of apoptotic death in epilepsy has been affirmed by other groups [65–67]. Some authors state that neuronal loss after epileptic seizures may be due to an active mechanism of apoptotic death, due to the finding that different classes of regulatory proteins in this process are activated by seizures, including caspases, death receptors, and proteins. The family of Bcl-2, DNA fragmentation pattern, and structural changes all predict apoptotic cell death (Graham [63,68–71]).

In summary, our results show the increase in immunodetection with annexin V and the existence of a correlation between the concentration of IL-6 and the percentage of annexin V positive neurons in patients with DRTLE, which speaks in favor of the occurrence of an apoptotic neuronal death process. Similarly, we have shown an increase in the expression of NF-kBas well as the cytokines IL-1β and IL-6, and immunodetection of annexin V co-expressed with positive synaptophysin cells. All these findings support the occurrence of inflammatory and apoptotic processes in DRTLE patients.

5. Conclusions

The evaluation of the participation of inflammatory processes in DRTLE is aimed at answering one basic question: Is inflammation part of the etiopathogenic processes in DRTLE or are the seizures the result of inflammatory disorders? In this sense, our results describe the clinical evidence that once the epileptogenic tissue has been resected with consequent elimination of seizures (or a significant reduction of them), the observed inflammation disappears. This finding supports the hypothesis that the inflammation could be mostly the consequence of epileptogenic processes. Similarly, this study supports the criterion of the involvement of inflammation and the activation of pathways such as NF-κB, and the final occurrence of death processes of the neuronal population is represented.

All these results highlight the necessity for new research about the role of inflammation and the immune response in the CNS, particularly in DRTLE, in order to achieve an understanding of the epileptogenic mechanism in this clinical entity and open up new possible immunomodulator treatments, in particular for those cases in which surgery is not a therapeutic alternative.

Acknowledgments: The authors would like to thank all the members of the epilepsy surgery program from the International Center for Neurological Restoration in Havana, Cuba; We wish to thank María Luisa Rodríguez Cordero for her held with the bibliography and to our reviewers for their helpful comments

Author Contributions: Lourdes Lorigados Pedre conceived and designed the experiments, produced and analyzed the results, and wrote the paper.Lilia M. Morales Chacón evaluated the patients and participated in the writing of the paper. Nancy Pavón Fuentes, Teresa Serrano Sánchez, María de los A. Robinson Agramonte, Mei-Li Díaz Hung and Rachel M. Cruz-Xenes executed the studies and the analysis of the results. Bárbara Estupiñán Díaz obtained and processed brain tissue. Margarita M. Báez Martín analyzed the results and revised the paper. Sandra Orozco-Suárez facilitated the confocal microscope and the realization of the immunohistochemical studies.

Conflicts of Interest: The authors declare no conflict of interest.

References

1. Thurman, D.J.; Beghi, E.; Begley, C.E.; Berg, A.T.; Buchhalter, J.R.; Ding, D.; Hesdorffer, D.C.; Hauser, W.A.; Kazis, L.; Kobau, R.; et al. Standards for epidemiologic studies and surveillance of epilepsy. *Epilepsia* **2011**, *52* (Suppl. 7), 2–26. [CrossRef] [PubMed]
2. Laxer, K.D.; Trinka, E.; Hirsch, L.J.; Cendes, F.; Langfitt, J.; Delanty, N.; Resnick, T.; Benbadis, S.R. The consequences of refractory epilepsy and its treatment. *Epilepsy Behav.* **2014**, *37*, 59–70. [CrossRef] [PubMed]
3. Walker, L.E.; Frigerio, F.; Ravizza, T.; Ricci, E.; Tse, K.; Jenkins, R.E.; Sills, G.J.; Jorgensen, A.; Porcu, L.; Thippeswamy, T.; et al. Molecular isoforms of high-mobility group box 1 are mechanistic biomarkers for epilepsy. *J. Clin. Investig.* **2017**, 92001. [CrossRef] [PubMed]
4. Scheffer, I.E.; Berkovic, S.; Capovilla, G.; Connolly, M.B.; French, J.; Guilhoto, L.; Hirsch, E.; Jain, S.; Mathern, G.W.; Moshe, S.L.; et al. Ilae classification of the epilepsies: Position paper of the ilae commission for classification and terminology. *Epilepsia* **2017**, *58*, 512–521. [CrossRef] [PubMed]
5. Vezzani, A.; Auvin, S.; Ravizza, T.; Aronica, E.; Aronica, E.; Ravizza, T.; Zurolo, E.; Vezzani, A. Glia-neuronal interactions in ictogenesis and epileptogenesis: Role of inflammatory mediators astrocyte immune responses in epilepsy. *Glia* **2012**, *60*, 1258–1268.
6. Morales-Chacon, L.M.; Alfredo Sanchez, C.C.; Minou Baez, M.M.; Rodriguez, R.R.; Lorigados, P.L.; Estupinan, D.B. Multimodal imaging in nonlesional medically intractable focal epilepsy. *Front Biosci. (Elite Ed.)* **2015**, *7*, 42–57. [CrossRef] [PubMed]
7. Fois, A.; Vascotto, M. Use of intravenous immunoglobulins in drug-resistant epilepsy. *Childs Nerv. Syst.* **1990**, *6*, 400–405. [CrossRef] [PubMed]
8. Aarli, J.A. Epilepsy and the immune system. *Arch. Neurol.* **2000**, *57*, 1689–1692. [CrossRef] [PubMed]
9. Ravizza, T.; Balosso, S.; Vezzani, A. Inflammation and prevention of epileptogenesis. *Neurosci. Lett.* **2011**, *497*, 223–230. [CrossRef] [PubMed]
10. Ozkara, C.; Vigevano, F. Immuno- and antiinflammatory therapies in epileptic disorders. *Epilepsia* **2011**, *52* (Suppl. 3), 45–51. [CrossRef] [PubMed]
11. Callenbach, P.M.; Jol-Van Der Zijde, C.M.; Geerts, A.T.; Arts, W.F.; Van Donselaar, C.A.; Peters, A.C.; Stroink, H.; Brouwer, O.F.; Van Tol, M.J. Immunoglobulins in children with epilepsy: The dutch study of epilepsy in childhood. *Clin. Exp. Immunol.* **2003**, *132*, 144–151. [CrossRef] [PubMed]
12. Beghi, E.; Shorvon, S. Antiepileptic drugs and the immune system. *Epilepsia* **2011**, *52* (Suppl. 3), 40–44. [CrossRef] [PubMed]
13. Vezzani, A.; Bartfai, T.; Bianchi, M.; Rossetti, C.; French, J. Therapeutic potential of new antiinflammatory drugs. *Epilepsia* **2011**, *52* (Suppl. 8), 67–69. [CrossRef] [PubMed]

14. Vezzani, A. Anti-inflammatory drugs in epilepsy: Does it impact epileptogenesis? *Expert. Opin. Drug Saf.* **2015**, *14*, 583–592. [CrossRef] [PubMed]
15. Lorigados-Pedre, L.; Morales-Chacon, L.; Pavon-Fuentes, N.; Serrano-Sanchez, T.; Robinson-Agramonte, M.A.; Garcia-Navarro, M.E.; der-del Busto, J.E. Immunological disorders in epileptic patients are associated to the epileptogenic focus localization. *Rev. Neurol.* **2004**, *39*, 101–104. [PubMed]
16. Lorigados, P.L.; Morales Chacon, L.M.; Orozco, S.S.; Pavon, F.N.; Estupinan, D.B.; Serrano, S.T.; Garcia, M.I.; Rocha, A.L. Inflammatory mediators in epilepsy. *Curr. Pharm. Des.* **2013**, *19*, 6766–6772. [CrossRef]
17. Algattas, H.; Huang, J.H. Traumatic brain injury pathophysiology and treatments: Early, intermediate, and late phases post-injury. *Int. J. Mol. Sci.* **2013**, *15*, 309–341. [CrossRef] [PubMed]
18. Gales, J.M.; Jehi, L.; Nowacki, A.; Prayson, R.A. The role of histopathologic subtype in the setting of hippocampal sclerosis-associated mesial temporal lobe epilepsy. *Hum. Pathol.* **2017**, *63*, 79–88. [CrossRef] [PubMed]
19. Van Vliet, E.A.; Aronica, E.; Vezzani, A.; Ravizza, T. Neuroinflammatory pathways as treatment targets and biomarker candidates in epilepsy: Emerging evidence from preclinical and clinical studies. *Neuropathol. Appl. Neurobiol.* **2017**, *4*. [CrossRef] [PubMed]
20. Yu, N.; Liu, H.; Di, Q. Modulation of immunity and the inflammatory response: A new target for treating drug-resistant epilepsy. *Curr. Neuropharmacol.* **2013**, *11*, 114–127. [PubMed]
21. Heida, J.G.; Moshe, S.L.; Pittman, Q.J. The role of interleukin-1beta in febrile seizures. *Brain Dev.* **2009**, *31*, 388–393. [CrossRef] [PubMed]
22. Ishikawa, N.; Kobayashi, Y.; Fujii, Y.; Kobayashi, M. Increased interleukin-6 and high-sensitivity c-reactive protein levels in pediatric epilepsy patients with frequent, refractory generalized motor seizures. *Seizure* **2015**, *25*, 136–140. [CrossRef] [PubMed]
23. Uludag, I.F.; Duksal, T.; Tiftikcioglu, B.I.; Zorlu, Y.; Ozkaya, F.; Kirkali, G. IL-1β, IL-6 and IL1Ra levels in temporal lobe epilepsy. *Seizure* **2015**, *26*, 22–25. [CrossRef] [PubMed]
24. Ravizza, T.; Noe, F.; Zardoni, D.; Vaghi, V.; Sifringer, M.; Vezzani, A. Interleukin converting enzyme inhibition impairs kindling epileptogenesis in rats by blocking astrocytic IL-1β production. *Neurobiol. Dis.* **2008**, *31*, 327–333. [CrossRef] [PubMed]
25. Vezzani, A.; Fujinami, R.S.; White, H.S.; Preux, P.M.; Blumcke, I.; Sander, J.W.; Loscher, W. Infections, inflammation and epilepsy. *Acta Neuropathol.* **2016**, *131*, 211–234. [CrossRef] [PubMed]
26. Rijkers, K.; Majoie, H.J.; Hoogland, G.; Kenis, G.; De, B.M.; Vles, J.S. The role of interleukin-1 in seizures and epilepsy: A critical review. *Exp. Neurol.* **2009**, *216*, 258–271. [CrossRef] [PubMed]
27. Engel, J., Jr. Update on surgical treatment of the epilepsies. Summary of the second international palm desert conference on the surgical treatment of the epilepsies (1992). *Neurology* **1993**, *43*, 1612–1617. [CrossRef] [PubMed]
28. Nagy, C.; Maheu, M.; Lopez, J.P.; Vaillancourt, K.; Cruceanu, C.; Gross, J.A.; Arnovitz, M.; Mechawar, N.; Turecki, G. Effects of postmortem interval on biomolecule integrity in the brain. *J. Neuropathol. Exp. Neurol.* **2015**, *74*, 459–469. [CrossRef] [PubMed]
29. Blair, J.A.; Wang, C.; Hernandez, D.; Siedlak, S.L.; Rodgers, M.S.; Achar, R.K.; Fahmy, L.M.; Torres, S.L.; Petersen, R.B.; Zhu, X.; et al. Individual case analysis of postmortem interval time on brain tissue preservation. *PLoS ONE* **2016**, *11*, e0151615.
30. Banuelos-Cabrera, I.; Cuellar-Herrera, M.; Velasco, A.L.; Velasco, F.; Alonso-Vanegas, M.; Carmona, F.; Guevara, R.; Arias-Montano, J.A.; Rocha, L. Pharmacoresistant temporal lobe epilepsy modifies histamine turnover and h3 receptor function in the human hippocampus and temporal neocortex. *Epilepsia* **2016**, *57*, e76–e80. [CrossRef] [PubMed]
31. Lowry, O.H.; Rosebrough, N.J.; Farr, A.L.; Randall, R.J. Protein measurement with the folin phenol reagent. *J. Biol. Chem.* **1951**, *193*, 265–275. [PubMed]
32. Bernardino, L.; Ferreira, R.; Cristovao, A.J.; Sales, F.; Malva, J.O. Inflammation and neurogenesis in temporal lobe epilepsy. *Curr. Drug Targets. CNS Neurol. Disord.* **2005**, *4*, 349–360. [CrossRef] [PubMed]
33. Aronica, E.; Ravizza, T.; Zurolo, E.; Vezzani, A. Astrocyte immune responses in epilepsy. *Glia* **2012**, *60*, 1258–1268. [CrossRef] [PubMed]
34. Ravizza, T.; Gagliardi, B.; Noe, F.; Boer, K.; Aronica, E.; Vezzani, A. Innate and adaptive immunity during epileptogenesis and spontaneous seizures: Evidence from experimental models and human temporal lobe epilepsy. *Neurobiol. Dis.* **2008**, *29*, 142–160. [CrossRef] [PubMed]

35. Vezzani, A.; Viviani, B. Neuromodulatory properties of inflammatory cytokines and their impact on neuronal excitability. *Neuropharmacology* **2015**, *96*, 70–82. [CrossRef] [PubMed]

36. Fabene, P.F.; Navarro, M.G.; Martinello, M.; Rossi, B.; Merigo, F.; Ottoboni, L.; Bach, S.; Angiari, S.; Benati, D.; Chakir, A.; et al. A role for leukocyte-endothelial adhesion mechanisms in epilepsy. *Nat. Med.* **2008**, *14*, 1377–1383. [CrossRef] [PubMed]

37. Vezzani, A.; Pascente, R.; Ravizza, T. Biomarkers of epileptogenesis: The focus on glia and cognitive dysfunctions. *Neurochem. Res.* **2017**, *42*, 2089–2098. [CrossRef] [PubMed]

38. Yu, N.; Di, Q.; Hu, Y.; Zhang, Y.F.; Su, L.Y.; Liu, X.H.; Li, L.C. A meta-analysis of pro-inflammatory cytokines in the plasma of epileptic patients with recent seizure. *Neurosci. Lett.* **2012**, *514*, 110–115. [CrossRef] [PubMed]

39. Aronica, E.; Crino, P.B. Inflammation in epilepsy: Clinical observations. *Epilepsia* **2011**, *52* (Suppl. 3), 26–32. [CrossRef] [PubMed]

40. Mintzer, S. Metabolic consequences of antiepileptic drugs. *Curr. Opin. Neurol.* **2010**, *23*, 164–169. [CrossRef] [PubMed]

41. Mintzer, S.; Miller, R.; Shah, K.; Chervoneva, I.; Nei, M.; Skidmore, C.; Sperling, M.R. Long-term effect of antiepileptic drug switch on serum lipids and c-reactive protein. *Epilepsy Behav.* **2016**, *58*, 127–132. [CrossRef] [PubMed]

42. Liimatainen, S.; Fallah, M.; Kharazmi, E.; Peltola, M.; Peltola, J. Interleukin-6 levels are increased in temporal lobe epilepsy but not in extra-temporal lobe epilepsy. *J. Neurol.* **2009**, *256*, 796–802. [CrossRef] [PubMed]

43. Lorigados, L.; Morales, L.; Pavón, N.; Serrano, T.; Robinson, M.A.; García, M.E.; Bender, J.E. Alteraciones inmunológicas en pacientes epilépticos asociadas a la localización del foco epileptogénico. *Rev. Neurol.* **2004**, *39*, 101–104.

44. Laflamme, N.; Rivest, S. Toll-like receptor 4: The missing link of the cerebral innate immune response triggered by circulating gram-negative bacterial cell wall components. *FASEB J.* **2001**, *15*, 155–163. [CrossRef] [PubMed]

45. Jansson, D.; Rustenhoven, J.; Feng, S.; Hurley, D.; Oldfield, R.L.; Bergin, P.S.; Mee, E.W.; Faull, R.L.; Dragunow, M. A role for human brain pericytes in neuroinflammation. *J. Neuroinflamm.* **2014**, *11*, 104–111. [CrossRef] [PubMed]

46. Librizzi, L.; Noe, F.; Vezzani, A.; De, C.M.; Ravizza, T. Seizure-induced brain-borne inflammation sustains seizure recurrence and blood-brain barrier damage. *Ann. Neurol.* **2012**, *72*, 82–90. [CrossRef] [PubMed]

47. Louboutin, J.P.; Strayer, D.S. Relationship between the chemokine receptor CCR5 and microglia in neurological disorders: Consequences of targeting CCR5 on neuroinflammation, neuronal death and regeneration in a model of epilepsy. *CNS Neurol. Disord. Drug Targets* **2013**, *12*, 815–829. [CrossRef] [PubMed]

48. Riazi, K.; Galic, M.A.; Pittman, Q.J. Contributions of peripheral inflammation to seizure susceptibility: Cytokines and brain excitability. *Epilepsy Res.* **2010**, *89*, 34–42. [CrossRef] [PubMed]

49. Cuellar-Herrera, M.; Velasco, A.L.; Velasco, F.; Chavez, L.; Orozco-Suarez, S.; Armagan, G.; Turunc, E.; Bojnik, E.; Yalcin, A.; Benyhe, S.; et al. Mu opioid receptor mrna expression, binding, and functional coupling to g-proteins in human epileptic hippocampus. *Hippocampus* **2012**, *22*, 122–127. [CrossRef] [PubMed]

50. Choi, J.; Nordli, D.R., Jr.; Alden, T.D.; DiPatri, A., Jr.; Laux, L.; Kelley, K.; Rosenow, J.; Schuele, S.U.; Rajaram, V.; Koh, S. Cellular injury and neuroinflammation in children with chronic intractable epilepsy. *J. Neuroinflamm.* **2009**, *6*, 38. [CrossRef] [PubMed]

51. Van Gassen, K.L.; de Wit, M.; Koerkamp, M.J.; Rensen, M.G.; van Rijen, P.C.; Holstege, F.C.; Lindhout, D.; de Graan, P.N. Possible role of the innate immunity in temporal lobe epilepsy. *Epilepsia* **2008**, *49*, 1055–1065. [CrossRef] [PubMed]

52. Turrin, N.P.; Rivest, S. Innate immune reaction in response to seizures: Implications for the neuropathology associated with epilepsy. *Neurobiol. Dis.* **2004**, *16*, 321–334. [CrossRef] [PubMed]

53. Gahring, L.C.; Days, E.L.; Kaasch, T.; Gonzalez de, M.M.; Owen, L.; Persiyanov, K.; Rogers, S.W. Pro-inflammatory cytokines modify neuronal nicotinic acetylcholine receptor assembly. *J. Neuroimmunol.* **2005**, *166*, 88–101. [CrossRef] [PubMed]

54. Eriksson, C.; Zou, L.P.; Ahlenius, S.; Winblad, B.; Schultzberg, M. Inhibition of kainic acid induced expression of interleukin-1 beta and interleukin-1 receptor antagonist mrna in the rat brain by nmda receptor antagonists. *Brain Res. Mol. Brain Res.* **2000**, *85*, 103–113. [CrossRef]

55. Vezzani, A.; Conti, M.; De, L.A.; Ravizza, T.; Moneta, D.; Marchesi, F.; De Simoni, M.G. Interleukin-1beta immunoreactivity and microglia are enhanced in the rat hippocampus by focal kainate application: Functional evidence for enhancement of electrographic seizures. *J. Neurosci.* **1999**, *19*, 5054–5065. [PubMed]

56. Lorigados, P.L.; Orozco, S.S.; Morales, C.L.; Garcia, M.I.; Estupinan, D.B.; Bender del Busto, J.E.; Pavon, F.N.; Paula, P.B.; Rocha, A.L. Neuronal death in the neocortex of drug resistant temporal lobe epilepsy patients. *Neurologia* **2008**, *23*, 555–565.

57. Henshall, D.C.; Engel, T. Contribution of apoptosis-associated signaling pathways to epileptogenesis: Lessons from Bcl-2 family knockouts. *Front Cell Neurosci.* **2013**, *7*. [CrossRef] [PubMed]

58. Crampton, S.J.; O'Keeffe, G.W. Nf-kappab: Emerging roles in hippocampal development and function. *Int. J. Biochem. Cell Biol.* **2013**, *45*, 1821–1824. [CrossRef] [PubMed]

59. Teocchi, M.A.; Ferreira, A.Ç.; da Luz de Oliveira, E.P.; Tedeschi, H.; D'Souza-Li, L. Hippocampal gene expression dysregulation of klotho, nuclear factor kappa b and tumor necrosis factor in temporal lobe epilepsy patients. *J. Neuroinflamm.* **2013**, *10*, 53. [CrossRef] [PubMed]

60. Crespel, A.; Coubes, P.; Rousset, M.C.; Brana, C.; Rougier, A.; Rondouin, G.; Bockaert, J.; Baldy-Moulinier, M.; Lerner-Natoli, M. Inflammatory reactions in human medial temporal lobe epilepsy with hippocampal sclerosis. *Brain Res.* **2002**, *952*, 159–169. [CrossRef]

61. DeFelipe-Oroquieta, J.; Arellano, J.I.; Alonso, L.; Muñoz, A. Neuropatología de la epilepsia del lóbulo temporal: Alteraciones primarias y secundarias de los circuitos corticales y epileptogenicidad. *Rev. Neurol.* **2002**, *34*, 401–408. [PubMed]

62. Lorigados, L.; Orozco, S.; Morales-Chacon, L.; Estupiñan, B.; García, I.; Rocha, L. Excitotoxicity and neuronal death in epilepsy. *Biotechnol. Apl.* **2013**, *30*, 9–16.

63. Liou, A.K.; Clark, R.S.; Henshall, D.C.; Yin, X.M.; Chen, J. To die or not to die for neurons in ischemia, traumatic brain injury and epilepsy: A review on the stress-activated signaling pathways and apoptotic pathways. *Prog. Neurobiol.* **2003**, *69*, 103–142. [CrossRef]

64. Rojas, J.; Gonzalez, M.E.; Lorigados, L.; Morales, L.; Riverón, G.; Bauz, Y. Oxidative estress markers in surgically treated patients with refractory epilepsy. *Clin. Biochem.* **2007**, *40*, 292–298.

65. Henshall, D.C. Apoptosis signalling pathways in seizure-induced neuronal death and epilepsy. *Biochem. Soc. Trans.* **2007**, *35*, 421–423. [CrossRef] [PubMed]

66. Henshall, D.C.; Simon, R.P. Epilepsy and apoptosis pathways. *J. Cereb. Blood Flow Metab.* **2005**, *25*, 1557–1572. [CrossRef] [PubMed]

67. Narkilahti, S.; Nissinen, J.; Pitkanen, A. Administration of caspase 3 inhibitor during and after status epilepticus in rat: Effect on neuronal damage and epileptogenesis. *Neuropharmacology* **2003**, *44*, 1068–1088. [CrossRef]

68. Graham, S.H.; Chen, J.; Stetler, R.A.; Zhu, R.L.; Jin, K.L.; Simon, R.P. Expression of the proto-oncogene bcl-2 is increased in the rat brain following kainate-induced seizures. *Restor. Neurol. Neurosci.* **1996**, *9*, 243–250. [PubMed]

69. Roy, M.; Hom, J.J.; Sapolsky, R.M. Hsv-mediated delivery of virally derived anti-apoptotic genes protects the rat hippocampus from damage following excitotoxicity, but not metabolic disruption. *Gene Ther.* **2002**, *9*, 214–219. [CrossRef] [PubMed]

70. Naziroglu, M.; Ovey, I.S. Involvement of apoptosis and calcium accumulation through trpv1 channels in neurobiology of epilepsy. *Neuroscience* **2015**, *293*, 55–66. [CrossRef] [PubMed]

71. Teocchi, M.A.; D'Souza-Li, L. Apoptosis through death receptors in temporal lobe epilepsy-associated hippocampal sclerosis. *Mediat. Inflamm.* **2016**, *2016*. [CrossRef] [PubMed]

behavioral sciences

Article

Long-Term Electroclinical and Employment Follow up in Temporal Lobe Epilepsy Surgery. A Cuban Comprehensive Epilepsy Surgery Program

Lilia Maria Morales Chacón *, Ivan Garcia Maeso, Margarita M. Baez Martin,
Juan E. Bender del Busto, María Eugenia García Navarro, Nelson Quintanal Cordero,
Bárbara Estupiñan Díaz, Lourdes Lorigados Pedre, Ricardo Valdés Yerena, Judith Gonzalez,
Randy Garbey Fernandez and Abel Sánchez Coroneux

Epilepsy Surgery Program International Center for Neurological Restoration, 25th Ave, No 15805, Havana, Cuba;
ivangarciamaeso@gmail.com (I.G.M.); minou@neuro.ciren.cu (M.M.B.M.); jebender@infomed.sld.cu (J.E.B.d.B.);
marugeniagarcian@gmail.com (M.E.G.N.); nquintanal@neuro.ciren.cu (N.Q.C.); baby@neuro.ciren.cu (B.E.D.);
lourdesl@neuro.ciren.cu (L.L.P.); rvaldes@neuro.ciren.cu (R.V.Y.); judith@neuro.ciren.cu (J.G.);
randis0770@gmail.com (R.G.F.); abel@neuro.ciren.cu (A.S.C.)
* Correspondence: lily@neuro.ciren.cu; Tel.: +53-72-730-920

Received: 26 December 2017; Accepted: 20 January 2018; Published: 01 February 2018

Abstract: The purpose of this paper is to present a long- term electroclinical and employment follow up in temporal lobe epilepsy (TLE) patients in a comprehensive epilepsy surgery program. Forty adult patients with pharmacoresistant TLE underwent detailed presurgical evaluation. Electroencephalogram (EEG) and clinical follow up assessment for each patient were carried out. The occurrence of interictal epileptiform activity (IEA) and absolute spike frequency (ASF) were tabulated before and after 1, 6, 12, 24 and 72 months surgical treatment. Employment status pre- to post-surgery at the last evaluated period was also examined. Engel scores follow-up was described as follows: at 12 months 70% (28) class I, 10% (4) class II and 19% (8) class III-IV; at 24 months after surgery 55.2% (21) of the patients were class I, 28.9% (11) class II and 15.1% (6) class III-IV. After one-year follow up 23 (57.7%) patients were seizure and aura-free (Engel class IA). These figures changed to 47.3%, and 48.6% respectively two and five years following surgery whereas 50% maintained this condition in the last follow up period. A decline in the ASF was observed from the first year until the sixth year after surgery in relation to the preoperative EEG. The ASF one year after surgery allowed to distinguish "satisfactory" from "unsatisfactory" seizure relief outcome at the last follow up. An adequate social functioning in terms of education and employment in more than 50% of the patients was also found. Results revealed the feasibility of conducting a successful epilepsy surgery program with favorable long term electroclinical and psychosocial functioning outcomes in a developing country as well.

Keywords: temporal lobe epilepsy; epilepsy surgery; long term follow up; Electroencephalogram interictal epileptiform discharge; employment

1. Introduction

Temporal lobe epilepsy (TLE) associated with mesial temporal sclerosis is considered the most common and pharmacoresistant type of epilepsy in adults; therefore, 70–80% of epilepsy surgeries are performed in the temporal lobe. That is the reason why our epilepsy surgery program precisely initiated with this epilepsy type. According to statistics available as of 2005 epilepsy prevalence in Cuba is approximately 3. 1/1000 people [1]. Thus, an estimated 80,000 people in the country suffer from epilepsy, about 24,000 are pharmarcoresistant, and approximately 2400 might have surgery indications.

However, the success of epilepsy surgery (ES) depends upon the early identification of potential surgical candidates based on the available resources and technologies [2]. Eventually, after temporal lobe resective epilepsy surgery, the patient is 70% likely to be seizure-free, and over 30% to be free of antiepileptic drug (AEDs) within 2 years after surgery [3,4].

Epilepsy surgery still remains the most underutilized of all acceptable medical intervention not only in developed countries but also in developing countries as the causes of surgical failure are not clearly understood, and the reasons for this are creating an enormous treatment gap [5].

On the other hand, there is little literature available on epilepsy surgery in developing nations and the access to and availability of epilepsy management programs are very limited. Thus, there are multiple social, economic, and medical challenges in establishing successful epilepsy surgery programs in low- and middle-income countries, and the issue of developing epilepsy centers in resource-limited areas in a large scale is essential [6–8].

Although the result of epilepsy surgery has improved over time [9–12], the few studies that have assessed chronological changes in surgical outcome in medial TLE have generally reported cross-sectional analyses limited to seizure outcomes in low- and middle-income countries. Additionally, the Engel and ILAE classification systems address only seizure outcome, and do not assess psychosocial, behavioral, cognitive and vocational development; all vital to gauge the utility of epilepsy surgery [13,14]. Even when considering both cross-sectional and longitudinal studies, there is insufficient research directly examining the longer-term educational and vocational outcomes of adult TLE surgery patients.

This article illustrates the results of a long- term follow up in TLE patients operated in the first comprehensive surgery program carried out at the International Center for Neurological Restoration in Havana, Cuba. A longitudinal Electroencephalogram (EEG), and clinical long term follow up for each patient was assessed. Social functioning outcomes are also addressed in this paper.

2. Materials and Methods

2.1. Patient Population

Epileptic patients with pharmacoresistant focal epilepsy were referred from different regions of the country. Cases were required to be non-responsive to at least 2 appropriate AEDs trials due to inefficacy and intolerance; hence recurrently compromised by seizures. Then, patients were consecutively admitted to the first comprehensive surgery program from May 2012 to September 2015. Only subjects submitted to TLE resection with over one-year follow-up after surgery were included whereas those with prior brain surgery were left out. Lastly, patients were evaluated by an epileptologist (LM) before being operated by an epilepsy surgeon (IG). Family and patient's informed consent was received in all cases.

2.2. Presurgical Evaluation

Each patient underwent noninvasive presurgical evaluation program including: (a) prolonged video-electroencephalography (VEEG) monitoring with scalp electrodes placed according to the international 10–20 system and additional anterior temporal electrodes; (b) Magnetic Resonance Imaging (MRI) scans with a 1.5 T scanner (Siemens Magnetom Symphony) including the following sequences: T1-weighted images with and without gadolinium-DTPA, T2-weighted images, fluid-attenuated inversion recovery images and magnetization-prepared rapid gradient echo sequences. Axial images were obtained with a modified angulation parallel to the temporal lobe long axis to assess the mesiotemporal structures; (c) A comprehensive battery of neuropsychological tests (attention assessment, memory, higher verbal and visual functions); (d) Perimetric evaluation and quadrant visual evoked potential VEPs [15].

Voxel based morphometric MRI post processing comprising volumetric analysis and functional neuroimaging using interictal and ictal brain single photon emission computed tomography and Magnetic Resonance Spectroscopy (MRS) were carried out in only 15% of patients when MRI was normal, and when there was discordance between VEEG and MRI, in accordance with our previously published protocol [16].

In the VEEG, ictal EEG patterns at seizure onset were categorized Type I, Type II A&B, Type III based on established criteria by Ebersole and Pacia [17]. With respect to the resection site ictal and interictal EEG findings were classified concordant (if 75% or more corresponded to the site of seizure origin (based on seizure semiology, MRI abnormality, and/or area resected) or discordant (i.e., any evidence of a wider, even if lateralized, spatial distribution outside the resection area involving more than one seizure. Ictal EEG activity was categorized localized to the presumed lobe of seizure origin, lateralized to the presumed hemisphere of seizure origin and diffuse (uncertain hemispheric origin). The distribution of interictal epileptiform discharges (IEDs) during prolonged video-EEG monitoring was assessed by analyzing 15 minutes interictal EEG samples every 1 h. The data recorded in relation to events identified by button presses or by seizure or spike detection programs was also reviewed. Patients underwent VEEG monitoring for 10.7 ± 3.14 days.

Ictal and interictal EEG was analyzed by a qualified epileptologist involved in the study (LM).

Patient test results were discussed in an epilepsy surgery conference including a multidisciplinary team.

2.3. Surgical Procedure and Resection Size

Anteromedial Temporal lobectomy tailored by Electrocorticography (ECoG) recording was carried out by a neurosurgeon (IG). ECoG data acquisition was performed with a Medicid-5 digital EEG system (Neuronic SA, Cuba) made in Cuba, using AD-TECH subdural electrodes (grid and strips).The superior gyrus was spared with the neocortical resection of the middle and inferior temporal gyrus. The extent of anterior temporal neocortical resection was adjusted according to the ECoG findings. The absolute longitude of lateral and mesial resected tissue was measured by the neurosurgeon (IG) using slices of images in T1, T2 and FLAIR six months after surgery. The lateral (neocortical) aspect included the distance (in mm) between the posterior edge of the internal table and the anterior limit of the resected area considering the middle temporal gyrus in the axial slices of MRI whereas the mesial aspect was calculated from axial slices in parallel with the preserved hippocampus. Two tangential lines were drawn: one between the posterior border of the resection and the contralateral hippocampus, and the other between the anterior limit of the preserved hippocampus and the side of resection. The distance between these tangential lines was the mesial longitude.

2.4. Tissue Characterization and Histopathological Examination

Resected specimens varied in size depending on the ECoG result. Haematoxylin-eosin and Kluver-Barrera myelin special stain were performed in mesial and neocortical specimens.

Hippocampal sclerosis (HS) was defined by neuronal loss in CA1, CA3 and CA4 regions of the hippocampus. Gliofibrillary acidic protein (GENNOVA, dilution 1/50) was used to qualitatively evaluate the astrogliosis as a consequence of the neuronal loss in the hippocampus and neocortex as well as the baloon cells. Synaptophysin (GENNOVA, ready to use) was performed when immunohistochemical staining was necessary.

The presence of focal cortical dysplasia (FCD) and HS was independently confirmed by two neuropathologists. For mycroscopic diagnosis, and FCD classification, the system proposed by the International League Against epilepsy was used [18]. For Central Nervous System tumor histopathological diagnosis purpose WHO classification was used [19].

2.5. Post Operative Follow-Up

Clinical follow up assessment for each patient was carried out six months (n = 40); one (n = 40), two (n = 38), five (n = 37) years; and later from six to fourteen years (n = 30) (mean 9.7 years) after surgery 42.5% patients were followed up for over ten years and 75% for at least seven years. Overall, the mean follow-up was 8.6 ± 3.9 years. Data were collected prospectively.

Postsurgical seizure outcome assessment was based on the system proposed by Engel. [Engel class I, free of disabling seizures; class IA, seizure-free; class II, rare seizures (fewer than three seizures per year); class III, worthwhile improvement (reduction in seizures of 80% or more); class IV, no benefit] [20]. To illustrate, and for some statistical analysis, class I was classified as "satisfactory" outcome, while classes II, III and IV as "unsatisfactory" seizure relief outcome.

Digital Scalp 30–60 min routine EEG recording was reviewed and interpreted by experienced electroencephalographers (LM) to assess the presence of interictal epileptiform discharges (IEDs) defined as sharp waves, spikes and electrographic seizure. IEDs were characterized as temporal (ipsilateral or contralateral to surgery), bitemporal, extratemporal, and generalized. Scalp EEG analysis was done using bipolar (longitudinal and transverse with temporal chains) and referential montages. The occurrence of IED, and absolute spike frequency (ASF) calculated as IED/min were then tabulated before and after 1, 6, 12, 24 and 72 months surgical treatment.

In order to investigate social functioning, the employment and educational status pre-to post-surgery, in the last follow up was analyzed.

2.6. Statistics Analysis

Data were collected from follow-up visits and sequentially entered into the database. These data were summarized with descriptive statistics for each variable comprising means, medians, and standard deviations for continuous variables and frequencies for categorical variables. Normality of the data was tested using Shapiro-Wilk test. Results showed non-normal distribution of some variables. Besides, non-parametric inference was used for comparisons. The Mann-Whitney U test was used to compare nonparametric values. The Friedman ANOVA and sign test were used to compare the electroclinical follow up one, six months; one, two, five years; and from six to fourteen years following surgery. Differences were considered statistically significant at the 5% level.

2.7. Ethical Considerations

All the procedures followed the rules of the Declaration of Helsinki of 1975 for human research, and the study was approved by the scientific and ethics committee from the International Center for Neurological Restoration (CIREN37/2012).

3. Results

3.1. Demographic Profile and Electroclinical Features

Patients undergoing noninvasive presurgical evaluation at the CIREN comprehensive epilepsy surgery program were carefully selected by our interdisciplinary team of epileptologists. During extracranial Video-EEG monitoring a mean of 13.1 ± 9.8 seizures per patient was recorded with a mean Video -EEG monitoring efficiency equal 0.99. In the whole group the first seizure occurred at day 3; and the third at day 5. Awake and sleep seizures indexes were 0.77 and 0.24 respectively. In 70% (28 of 40) of cases the antiepileptic drug regimen was partially reduced during the video-EEG session. Type I Ictal EEG pattern was documented in 60% of cases; and 62.5% had bitemporal with unilateral predominance IED (See Table 1).

Table 1. Demographic, clinical and surgical cohort characteristics (40 patients).

Mean age at surgery (years ± SD range)	33.5 ± 9.7, (range 16–58)
Mean age at seizure onset (years ± SD range)	13.7 ± 11.3, (range 9–52)
Gender	Male: 22 (55%) Female: 18 (45%)
Mean epilepsy duration (year ± SD range)	19.6 ± 10.18 (range 2–42)
Precipitant event *n* (%)	31 (77.5%), febrile seizures 47.6%
Mean number of antiepileptic drugs tried ± SD (range)	5.95 ± 2.02, (range 3–10)

Table 1. *Cont.*

Resection lateralization *n* (%)	Right 19 (47.5%) Left 21 (52.5%)
Mean follow-up (month range) (year ± SD range)	8.6 ± 3.9 years (range 1–14)
Seizure types	Complex partial seizures: 100% Simple partial seizures: 76% History of secondary generalized tonic-clonic seizures: 84%
Ictal EEG pattern *n* (%)	Type I (5–9 Hz) 24 (60%) Type II (2–5 Hz) 16 (40%)
Ictal EEG topography, *n* (%)	Concordant 31 (77.4%) Discordant 8 (22.5%)
Preoperative interictal EEG, *n* (%)	Unilateral/concordant 15 (37.5%); Bilateral with ipsilateral predominance 5:1, 25 (62.5%)
Left hemisphere resection longitude (mean ± SD), Right hemisphere resection longitude (mean ± SD)	mesial 18.23 ± 6.76 mm, neocortical 41.6 ± 9.8 mm mesial 15.06 ± 5.19 mm, neocortical 40.19 ± 12.07

SD Standard deviation.

3.1.1. Temporal Lobectomy and Complications

The resection amount in the 40 patients submitted to anterior temporal lobectomy was based on a result combination obtained from presurgical evaluation and intraoperative ECoG findings Table 1. There were no significant differences in the resection size between right and left temporal lobectomies. Mann Whitney U test for neocortical $p = 0.46$, for mesial $p = 0.18$. COMPLICATIONS One case experienced dysphasia and right hemiparesis. Aseptic meningitis was treated postoperatively in two patients. Another patient suffered from bacterial meningitis. Quadrantanopsia was not considered as complication. One patient, categorized in Engel Class II, died from an orthopedic surgery four years after surgery.

3.1.2. Pathology

The most common etiology was Focal Cortical Dysplasia (FCD) associated with a principal lesion (FCD type III). 67.5% had FCD type IIIa (cortical lamination abnormalities in the temporal lobe associated with hippocampal sclerosis); 10% FCD type IIIb (cortical lamination abnormalities adjacent to a glial or glioneuronal tumor; one patient with FCD IIIc (cortical lamination abnormalities adjacent to vascular malformation, and six with only HS. Dual pathology was documented to be associated with pylocitic astrocytom and arachnoid cystic in two of the patients.

3.2. Electroclinical Follow Up

After one- year follow up, 23 (57.7%) were completely seizure free and aura free (Engel class IA) while two and five years after surgery the percentage changed from 47.3%, to 48.6% respectively. In the last follow up period 50% of the patients maintained this condition. However, there was no significant difference in the number of patient's seizure and aura free over the long time. Patients kept their AEDs for at least 2 years postsurgery.

Engel scores follow-up was described as follows: at 12 months 70% (28) class I, 10% (4) class II and 19% (8) class III-IV; at 24 months: 55.2% (21) of cases were class I, 28.9%(11) class II and 15.1% (6) class III-IV. Five years after surgery, 54.05% (20) class I, 35.1% (13) class II, 10.8% (4) class III. At the last follow-up period 55.1% (16) class I, 24.1% (7) class II and 20.6% (6) class III. There was a notable difference between clinical evolutions considering all evaluated period (Friedman ANOVA $p = 0.01286$, The percentage of patients in Engel class I decreased two years postsurgery in relation to the previous year ($p = 0.01$ Sign test). Overall, there was no substantial variation for Engel class I within 24 months and the last follow up period ($p > 0.05$ Sign test) (See Figure 1 and Table 2).

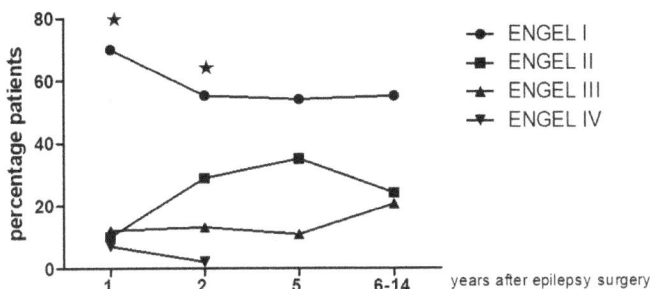

Figure 1. Long term clinical follow up using Engel Scale in temporal lobe epilepsy patients submitted to epilepsy surgery. Notice that the percentage of patients in Engel class I decreased two years postsurgery in relation to the previous year ($\star$ $p = 0.01$, Friedman ANOVA and Sign test). There was no substantial variation for Engel class I within 24 months and the last follow up period ($p > 0.05$).

Table 2. Year-by-year Clinical follow up by Engel class.

Follow-Up	Class I Patients *n*, (%)	Class II Patients *n*, (%)	Class III Patients *n*, (%)	Class IV Patients *n*, (%)
1 year, *n* = 40	28, (70%)	4, (10%)	5, (12%)	3 (7%)
2 year, *n* = 38	21, (55.2%)	11, (28.9)	5, (13.1%)	1 (2%)
5 year, *n* = 37	20, (54.05)	13, (35.1%)	4, (10.8%)	
6–14 year mean 9.7y *n* = 29	16, (55.1)	7, (24.1%)	6, (20.6%)	

Interictal Epileptiform Discharges on Post-Operative EEG

There were noteworthy changes in the ASF follow up (Friedman ANOVA $p = 0.0108$. A decline in the ASF was observed one, two and six years after surgery in relation to the preoperative EEG (p 0.003 Sign test). However, there were no differences in the ASF between the first year following surgery and the other evaluated periods. See Figure 2. Interestingly the ASF in the EEG recorded one year postsurgery was significantly different in "satisfactory" outcome cases from those with "unsatisfactory" seizure relief outcome $p = 0.02$ at the last clinical follow up Figure 3.

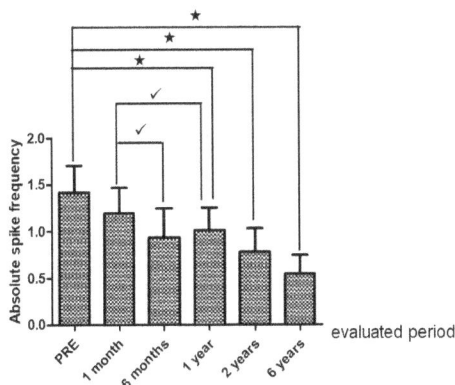

Figure 2. Absolute spike frequency (ASP) on pre-and postoperative Electroencephalogram EEG (one, six months and one, two and six year). Comparisons of pre-EEG with one, two and six years after epilepsy surgery ($\star$ $p < 0.05$). Comparisons between one month postsurgical EEG and six months and one year post surgery ($\sqrt{}$ $p < 0.05$) Friedman ANOVA and Sigh test. There were no differences in the SAF between the first year after and the other evaluated periods. Notice x axis: evaluated EFG period, y axis: mean and SEM of the ASF (spike/min).

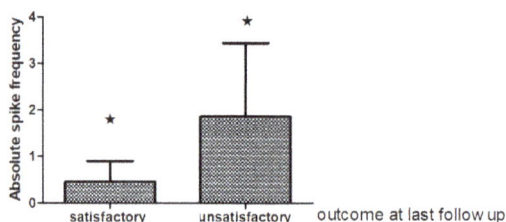

Figure 3. Bar grah showing comparisons of absolute spike frequency on the EEG one year postsurgery (mean and standart deviation SD) in temporal lobe patients with satisfactory (Engel class I) and unsatisfactory (Engel class II–IV) outcome at the last follow up (6–14 years) ($\bigstar$ $p < 0.05$, Mann Whitney U test).

3.3. Pre and Post-Surgery Education and Employment Status in TLE Patients

In terms of education and employment status, 41.6% of TLE operated patients who were employed before surgery remained in regular work. 5% of patients moved to supported work and 7.8% began to study after surgery. A total of 27.7% of the patients remained unemployment whereas 13.8% became unemployed post-surgery. More than 70% of patients in both employment and unemployment group were seizure free. As a whole, over 50% of our patients showed an adequate social functioning in terms of education and employment fourteen years after TLE surgery.

4. Discussion

This study indicates a steady number of patients as seizure-free in a long-term temporal lobe epilepsy surgery outcome, and highlights the value of longitudinal postoperative EEG in epilepsy surgery follow up. Our results also provide evidence of an adequate social functioning in terms of education and employment in TLE operated patients.

In TLE patients undergoing surgery Engel class I was achieved in 70% at 12 months' follow-up, 55.2%, 54.05% and 55.1% at 2, 5, and from 6 to14 years respectively. This representsa seizure decrease during the first year after surgery which remained stable from the second year until the last evaluated period. In agreement with our results, other studies have shown seizure free in the range of 60–70% after temporal lobectomy [21–23]. The proportion of patients free of seizure and aura over the long time in the current study (57.7%) was similar compared to a recent report that also showed complete seizure freedom in 65% of patients at 1year and in 56.5% at long-term follow-up of $\geq$5 years after temporal lobectomy [24]. Another author that included temporal and extratemporal resections detailed that >80% of the patients experienced class I outcome at the last four-year follow-up [25].

Our results are in keeping with other series from developing countries such as Argentina which found Engel class I outcome in 68.21% at 12 months' follow-up [26]. Mikati MA also stated 70% Engel class I, 9% class II, 14% class III, and 7% class IV after resective surgery in 93 adults and children who had undergone epilepsy surgery involving temporal resections in 54% of the cases at the American University of Beirut [27].

It should be observed that our seizure freedom outcome was better than that reported by other developing countries such as india and Uganda. In India, for instance, excellent seizure outcome (seizure-free or having only auras) was achieved in 7/17 patients (41%) [28] and in Uganda's series 6/10 60% of patients were seizure-free after temporal lobe epilepsy surgery [29]. Remarkably, Mrabet KH, and Campos et al., reported 100% and 88.24% of patients in Engel class I after epilepsy surgery [30,31]. The former included 15 patients with hippocampal sclerosis in an epilepsy surgery program in Tunisia with French collaboration, whereas the latter comprised 17 cases. A lower number of operated epilepsy patients was reported in all these series. Our results are also comparable with those described in a larger series of 87 children with temporal lobe epilepsy with HS and lesion-related epilepsies with a non-invasive protocol [32].

An interesting finding in this research is that patients undergoing temporal resection experienced significantly favorable long-term outcome. Two of the few studies that have reported longitudinal follow-up during 10 years include a retrospective single-center study in 325 patients (adults and children), in which 48% were continuously seizure-free after five years and 41% after 10 years [33]. The second study reported 55% seizure-free (without or with auras) five years following surgery and 49% 10 years later in 615 adults, 497 with temporal lobe resection [34].

Clearly, such comparisons are limited by selection criteria and referral patterns, which are likely to differ from different centers in Latin American countries. In order to standardize these criteria, our cases were discussed in an epilepsy surgery conference.

Another remarkable fact in our patients was that the epilepsy duration continued for more than 10 years (mean 19 years); and that the number of AED treated was higher than five (3–10). Moreover, most patients had bitemporal IED with unilateral predominance in the preoperative EEG. On the other hand, the most common etiology was FCD associated with a principal lesion (FCD type III). Cortical dysplasia has been documented in histologic specimens removed for treatment of drug resistant TLE epilepsy and there is evidence that a standard anterior temporal lobectomy offers a good seizure outcome [35–37]. Corresponding results were 64.2% and 56.8% at 1 and ≥5 years, respectively in patients with isolated mesio temporal lesion, and 66.4% and 56.0%, respectively in patients with mesial TLE and additional FCD [38]. In recent years, FCD has been identified as a major cause of pharmacoresistant focal epilepsy in patients undergoing surgical resection [39,40].The rate of seizure free after resection changed from 52 to 68.9% [41–43].

Even with this electroclinical profile, our seizure freedom outcome (Engel class I) was equivalent to other series in developed and developing countries, which points to epilepsy surgery as a long term effective treatment for carefully selected patients with pharmacoresistant temporal lobe epilepsy. It is important to highlight that distinction between this study and other series are limited not only because of the use of the classification proposed by the Task Force of the ILAE but also due to the epilepsy type addressed in this research. The current ILAE classification encompasses the FCD Type III that includes FCD associated with other principal lesions based on previous reports of lesions showing dysplastic changes in histology after resection [44].

In relation to preoperative video EEG it is important to notice that in 60% of the cases, type I ictal EEG pattern was recognized at seizure onset. This pattern also anticipated a favorable clinical evolution five years postsurgery; and has been associated with good outcomes in patients with TLE in earlier studies [28,45]. Nowadays, the controversy about the relative contributions of ictal scalp VEEG, routine scalp outpatient interictal EEG, intracranial EEG and MRI for predicting seizure-free outcomes after temporal lobectomy still remains [24].

Regarding post-operative EEG follow-up, a decline in the ASF was observed one, two and six years after surgery in relation to the preoperative EEG. It should be mentioned that quantitative measures of changes have been applied in very few studies; and the spike frequency in postoperative EEGs has been evaluated with controversial results by other authors [46–49]. Gropel reported that the spike frequency decreased in fourteen patients with mesial TLE-HS at four months, one and two years after surgical treatment increased in one; but it did not change in seven [48].

It ihas been stated that longer duration EEGs or repeated EEGs performed at different time intervals following surgery might have increased the chance of capturing IED, especially if epochs of sleep were involved. An advantage of this study is that we were able to quantify the IED in different follow up periods.

We also found that the ASF in the EEG recorded one year postsurgery was significantly different in "satisfactory" outcome cases from those with "unsatisfactory" seizure relief outcome. As a whole, studies are conflicting as to whether interictal EEG findings predict seizure recurrence [23,46,47,50,51]. Some of them showed a strong predictive value [47,52], while others revealed no predictive value [48,53–55]. Conflicting results are feasible as populations examined varied considerably among studies in the majority of series focusing on anterior temporal lobectomy. Also, there is no standard time to carry out a post-operative EEG, and centers that perform it routinely set their own protocols [49,56].

The assessment of surgical outcome in epilepsy beyond seizure control has received little attention in terms of social functioning, and the studies investigating employment and education in adults are scarce. Moreover, the range of postoperative psychosocial adjustment issues has been well documented, particularly within the first 24 months after surgery.

We found an adequate social functioning in terms of education and employment fourteen years after TLE surgery in more than 50% of the patients. In our study, 41.6% of TLE operated patients who were employed before surgery remained in regular work. 5% of patients moved to supported work and 7.8% began to study after surgery.

Although most studies have investigated employment outcomes alongside measures of quality of life, [57–59], a minority of them has focused on employment status pre-and post-surgery [60–63]. Overall, there is a mix of improvements and reductions in occupational status [64,65]. In the majority of studies an improve of vocational outcomes has been suggested [60,66–70]. Other authors as Asztely and colleagues reported a decline in the number of patients employed full-time following surgery [71]. Reid K and cols demonstrated that employment outcomes are directly relevant to patient satisfaction with surgery up to ten years later [70]. In support of the surgical treatment efficacy some studies comparing surgery to ongoing medical management demonstrate a trend towards higher employment postsurgery [62,72–74]. One study that used a healthy control demonstrated that up to 10 years post-surgery, even though patients who were seizure free were more likely to be employed than those with recurrent seizures. Overall, the number of patients that were still working was notably less than healthy controls (61% compared to 96%) [75].

It is not surprising, that in adults deemed eligible for epilepsy surgery, employment is a commonly cited reason for electing to undergo surgery. Both patients and their families identify educational and vocational outcomes as important, with expectations of improvements post-surgery [76–79].

The principal constraint of this study is the low number of patients that precluded the extraction of valuable information about potential prognostic factors in seizure recurrence so that more cases should be recruited to address this issue. Nonetheless, our results revealed a steady number of patients being seizure-free in a long-term temporal lobe epilepsy surgery outcome, and highlights the value of longitudinal postoperative EEG in epilepsy surgery follow up. Equally, results confirm the possibility of conducting a successful epilepsy surgery program with favorable long term electroclinical and psychosocial functioning outcomes in a developing country as well.

Acknowledgments: The authors would like to sincerely thank all the members of the epilepsy surgery program from the International Center for Neurological Restoration in Havana, Cuba; especially the telemetry unit nurses as well as the clinical neurophysiology technicians for their collaboration and support. We wish to thank Odalys Morales Chacon for revising the English in this manuscript. We are also grateful to our reviewers for their helpful comments.

Author Contributions: All authors contributed substantially to the work reported and participated in the epilepsy surgery program. Lilia Morales Chacon participated in the design of the project, analysis, and discussion of the results, and wrote the manuscript.

Conflicts of Interest: The authors declare no conflict of interest.

References

1. Gonzalez-Pal, S.; Quintana, M.J.; Roman Lopez, J.R.; Fernandez-Perez, J.E. The direct cost of epilepsy in Cuba. A study in outpatients. *Rev. Neurol.* **2005**, *41*, 379–381. [PubMed]
2. Asadi-Pooya, A.A.; Sperling, M.R. Strategies for surgical treatment of epilepsies in developing countries. *Epilepsia* **2008**, *49*, 381–385. [CrossRef] [PubMed]
3. Tellez-Zenteno, J.F.; Ladino, L.D. Temporal epilepsy: Clinical, diagnostic and therapeutic aspects. *Rev. Neurol.* **2013**, *56*, 229–242. [PubMed]

4. Kwon, C.S.; Neal, J.; Tellez-Zenteno, J.; Metcalfe, A.; Fitzgerald, K.; Hernandez-Ronquillo, L.; Hader, W.; Wiebe, S.; Jette, N. Resective focal epilepsy surgery—Has selection of candidates changed? A systematic review. *Epilepsy Res.* **2016**, *122*, 37–43. [CrossRef] [PubMed]
5. Engel, J., Jr. What can we do for people with drug-resistant epilepsy? The 2016 Wartenberg Lecture. *Neurology* **2016**, *87*, 2483–2489. [CrossRef] [PubMed]
6. Kasradze, S.; Alkhidze, M.; Lomidze, G.; Japaridze, G.; Tsiskaridze, A.; Zangaladze, A. Perspectives of epilepsy surgery in resource-poor countries: A study in Georgia. *Acta Neurochir. (Wien)* **2015**, *157*, 1533–1540. [CrossRef] [PubMed]
7. Radhakrishnan, K. Challenges in the management of epilepsy in resource-poor countries. *Nat. Rev. Neurol.* **2009**, *5*, 323–330. [CrossRef] [PubMed]
8. Rathore, C.; Rao, M.B.; Radhakrishnan, K. National epilepsy surgery program: Realistic goals and pragmatic solutions. *Neurol. India* **2014**, *62*, 124–129. [PubMed]
9. Asadi-Pooya, A.A.; Rakei, S.M.; Kamgarpour, A.; Taghipour, M.; Ashjazadeh, N.; Razmkon, A.; Zare, Z.; Bagheri, M.H. Outcome after temporal lobectomy in patients with medically-refractory mesial temporal epilepsy in Iran. *J. Neurosurg. Sci.* **2017**, *61*, 277–282. [CrossRef] [PubMed]
10. Scott, R.A.; Lhatoo, S.D.; Sander, J.W. The treatment of epilepsy in developing countries: Where do we go from here? *Bull. World Health Organ.* **2001**, *79*, 344–351. [PubMed]
11. Mani, K.S.; Subbakrishna, D.K. Perspectives from a developing nation with special reference to rural areas. *Epilepsia* **2003**, *44* (Suppl. 1), 55–57. [CrossRef] [PubMed]
12. Mbuba, C.K.; Newton, C.R. Packages of care for epilepsy in low- and middle-income countries. *PLoS Med.* **2009**, *6*, e1000162. [CrossRef] [PubMed]
13. Cross, J.H.; Jayakar, P.; Nordli, D.; Delalande, O.; Duchowny, M.; Wieser, H.G.; Guerrini, R.; Mathern, G.W. Proposed criteria for referral and evaluation of children for epilepsy surgery: Recommendations of the Subcommission for Pediatric Epilepsy Surgery. *Epilepsia* **2006**, *47*, 952–959. [CrossRef] [PubMed]
14. Perry, M.S.; Duchowny, M. Surgical versus medical treatment for refractory epilepsy: Outcomes beyond seizure control. *Epilepsia* **2013**, *54*, 2060–2070. [CrossRef] [PubMed]
15. Morales, L.M.; Sanchez, C.; Bender, J.E.; Bosch, J.; Garcia, M.E.; Garcia, I.; Lorigados, L.; Estupinan, B.; Trapaga, O.; Baez, M.; et al. A neurofunctional evaluation strategy for presurgical selection of temporal lobe epilepsy patients. *MEDICC Rev.* **2009**, *11*, 29–35. [PubMed]
16. Morales-Chacón, L.M.; Alfredo Sanchez, C.C.; Minou Baez, M.M.; Rodriguez, R.R.; Lorigados, P.L.; Estupiñan, D.B. Multimodal imaging in nonlesional medically intractable focal epilepsy. *Front. Biosci.* **2015**, *7*, 42–57. [CrossRef]
17. Ebersole, J.S.; Pacia, S.V. Localization of temporal lobe foci by ictal EEG patterns. *Epilepsia* **1996**, *37*, 386–399. [CrossRef] [PubMed]
18. Blumcke, I.; Muhlebner, A. Neuropathological work-up of focal cortical dysplasias using the new ILAE consensus classification system—Practical guideline article invited by the Euro-CNS Research Committee. *Clin. Neuropathol.* **2011**, *30*, 164–177. [CrossRef] [PubMed]
19. Louis, D.N.; Perry, A.; Reifenberger, G.; Von Deimling, A.; Figarella-Branger, D.; Cavenee, W.K.; Ohgaki, H.; Wiestler, O.D.; Kleihues, P.; Ellison, D.W. The 2016 World Health Organization Classification of Tumors of the Central Nervous System: A summary. *Acta Neuropathol.* **2016**, *131*, 803–820. [CrossRef] [PubMed]
20. Engel, J., Jr. Update on surgical treatment of the epilepsies. Summary of the Second International Palm Desert Conference on the Surgical Treatment of the Epilepsies (1992). *Neurology* **1993**, *43*, 1612–1617. [CrossRef] [PubMed]
21. Engel, J., Jr.; McDermott, M.P.; Wiebe, S.; Langfitt, J.T.; Stern, J.M.; Dewar, S.; Sperling, M.R.; Gardiner, I.; Erba, G.; Fried, I.; et al. Early surgical therapy for drug-resistant temporal lobe epilepsy: a randomized trial. *JAMA* **2012**, *307*, 922–930. [CrossRef] [PubMed]
22. Tellez-Zenteno, J.F.; Dhar, R.; Wiebe, S. Long-term seizure outcomes following epilepsy surgery: A systematic review and meta-analysis. *Brain* **2005**, *128*, 1188–1198. [CrossRef] [PubMed]
23. Tonini, C.; Beghi, E.; Berg, A.T.; Bogliun, G.; Giordano, L.; Newton, R.W.; Tetto, A.; Vitelli, E.; Vitezic, D.; Wiebe, S. Predictors of epilepsy surgery outcome: A meta-analysis. *Epilepsy Res.* **2004**, *62*, 75–87. [CrossRef] [PubMed]

24. Goldenholz, D.M.; Jow, A.; Khan, O.I.; Bagic, A.; Sato, S.; Auh, S.; Kufta, C.; Inati, S.; Theodore, W.H. Preoperative prediction of temporal lobe epilepsy surgery outcome. *Epilepsy Res.* **2016**, *127*, 331–338. [CrossRef] [PubMed]

25. Chen, H.; Modur, P.N.; Barot, N.; Van Ness, P.C.; Agostini, M.A.; Ding, K.; Gupta, P.; Hays, R.; Mickey, B. Predictors of Postoperative Seizure Recurrence: A Longitudinal Study of Temporal and Extratemporal Resections. *Epilepsy Res. Treat.* **2016**, *2016*, 7982494. [CrossRef] [PubMed]

26. Donadio, M.; D'Giano, C.; Moussalli, M.; Barrios, L.; Ugarnes, G.; Segalovich, M.; Pociecha, J.; Vazquez, C.; Petre, C.; Pomata, H. Epilepsy surgery in Argentina: Long-term results in a comprehensive epilepsy centre. *Seizure* **2011**, *20*, 442–445. [CrossRef] [PubMed]

27. Mikati, M.A.; Ataya, N.; El-Ferezli, J.; Shamseddine, A.; Rahi, A.; Herlopian, A.; Kurdi, R.; Bhar, S.; Hani, A.; Comair, Y.G. Epilepsy surgery in a developing country (Lebanon): Ten years experience and predictors of outcome. *Epileptic Disord.* **2012**, *14*, 267–274. [PubMed]

28. Sylaja, P.N.; Radhakrishnan, K.; Kesavadas, C.; Sarma, P.S. Seizure outcome after anterior temporal lobectomy and its predictors in patients with apparent temporal lobe epilepsy and normal MRI. *Epilepsia* **2004**, *45*, 803–808. [CrossRef] [PubMed]

29. Boling, W.; Palade, A.; Wabulya, A.; Longoni, N.; Warf, B.; Nestor, S.; Alpitsis, R.; Bittar, R.; Howard, C.; Andermann, F. Surgery for pharmacoresistant epilepsy in the developing world: A pilot study. *Epilepsia* **2009**, *50*, 1256–1261. [CrossRef] [PubMed]

30. Mrabet, K.H.; Khemiri, E.; Parain, D.; Hattab, N.; Proust, F.; Mrabet, A. Epilepsy surgery program in Tunisia: An example of a Tunisian French collaboration. *Seizure* **2010**, *19*, 74–78. [CrossRef] [PubMed]

31. Campos, M.G. Epilepsy surgery in developing countries. *Handb. Clin. Neurol.* **2012**, *108*, 943–953. [CrossRef] [PubMed]

32. Jayalakshmi, S.; Panigrahi, M.; Kulkarni, D.K.; Uppin, M.; Somayajula, S.; Challa, S. Outcome of epilepsy surgery in children after evaluation with non-invasive protocol. *Neurol. India* **2011**, *59*, 30–36. [CrossRef] [PubMed]

33. De Tisi, J.; Bell, G.S.; Peacock, J.L.; McEvoy, A.W.; Harkness, W.F.; Sander, J.W.; Duncan, J.S. The long-term outcome of adult epilepsy surgery, patterns of seizure remission, and relapse: A cohort study. *Lancet* **2011**, *378*, 1388–1395. [CrossRef]

34. McIntosh, A.M.; Kalnins, R.M.; Mitchell, L.A.; Fabinyi, G.C.; Briellmann, R.S.; Berkovic, S.F. Temporal lobectomy: Long-term seizure outcome, late recurrence and risks for seizure recurrence. *Brain* **2004**, *127*, 2018–2030. [CrossRef] [PubMed]

35. Widdess-Walsh, P.; Kellinghaus, C.; Jeha, L.; Kotagal, P.; Prayson, R.; Bingaman, W.; Najm, I.M. Electro-clinical and imaging characteristics of focal cortical dysplasia: Correlation with pathological subtypes. *Epilepsy Res.* **2005**, *67*, 25–33. [CrossRef] [PubMed]

36. Srikijvilaikul, T.; Najm, I.M.; Hovinga, C.A.; Prayson, R.A.; Gonzalez-Martinez, J.; Bingaman, W.E. Seizure outcome after temporal lobectomy in temporal lobe cortical dysplasia. *Epilepsia* **2003**, *44*, 1420–1424. [CrossRef] [PubMed]

37. Santos, M.V.; de Oliveira, R.S.; Machado, H.R. Approach to cortical dysplasia associated with glial and glioneuronal tumors (FCD type IIIb). *Child's Nerv. Syst.* **2014**, *30*, 1869–1874. [CrossRef] [PubMed]

38. Schmeiser, B.; Hammen, T.; Steinhoff, B.J.; Zentner, J.; Schulze-Bonhage, A. Long-term outcome characteristics in mesial temporal lobe epilepsy with and without associated cortical dysplasia. *Epilepsy Res.* **2016**, *126*, 147–156. [CrossRef] [PubMed]

39. Harvey, A.S.; Cross, J.H.; Shinnar, S.; Mathern, B.W. Defining the spectrum of international practice in pediatric epilepsy surgery patients. *Epilepsia* **2008**, *49*, 146–155. [CrossRef] [PubMed]

40. Lerner, J.T.; Salamon, N.; Hauptman, J.S.; Velasco, T.R.; Hemb, M.; Wu, J.Y.; Sankar, R.; Donald, S.W.; Engel, J., Jr.; Fried, I.; et al. Assessment and surgical outcomes for mild type I and severe type II cortical dysplasia: A critical review and the UCLA experience. *Epilepsia* **2009**, *50*, 1310–1335. [CrossRef] [PubMed]

41. Xue, H.; Cai, L.; Dong, S.; Li, Y. Clinical characteristics and post-surgical outcomes of focal cortical dysplasia subtypes. *J. Clin. Neurosci.* **2016**, *23*, 68–72. [CrossRef] [PubMed]

42. Fauser, S.; Bast, T.; Altenmuller, D.M.; Schulte-Monting, J.; Strobl, K.; Steinhoff, B.J.; Zentner, J.; Schulze-Bonhage, A. Factors influencing surgical outcome in patients with focal cortical dysplasia. *J. Neurol. Neurosurg. Psychiatry* **2008**, *79*, 103–105. [CrossRef] [PubMed]

43. Fauser, S.; Zentner, J. Management of cortical dysplasia in epilepsy. *Adv. Tech. Stand. Neurosurg.* **2012**, *38*, 137–163. [PubMed]
44. Fauser, S.; Schulze-Bonhage, A. Epileptogenicity of cortical dysplasia in temporal lobe dual pathology: An electrophysiological study with invasive recordings. *Brain* **2006**, *129*, 82–95. [CrossRef] [PubMed]
45. Holmes, M.D.; Kutsy, R.L.; Ojemann, G.A.; Wilensky, A.J.; Ojemann, L.M. Interictal, unifocal spikes in refractory extratemporal epilepsy predict ictal origin and postsurgical outcome. *Clin. Neurophysiol.* **2000**, *111*, 1802–1808. [CrossRef]
46. Di, G.G.; Quarato, P.P.; Sebastiano, F.; Esposito, V.; Onorati, P.; Mascia, A.; Romanelli, P.; Grammaldo, L.G.; Falco, C.; Scoppetta, C.; et al. Postoperative EEG and seizure outcome in temporal lobe epilepsy surgery. *Clin. Neurophysiol.* **2004**, *115*, 1212–1219.
47. Tuunainen, A.; Nousiainen, U.; Mervaala, E.; Pilke, A.; Vapalahti, M.; Leinonen, E.; Paljarvi, L.; Riekkinen, P. Postoperative EEG and electrocorticography: Relation to clinical outcome in patients with temporal lobe surgery. *Epilepsia* **1994**, *35*, 1165–1173. [CrossRef] [PubMed]
48. Groppel, G.; Aull-Watschinger, S.; Baumgartner, C. Temporal evolution and prognostic significance of postoperative spikes after selective amygdala-hippocampectomy. *J. Clin. Neurophysiol.* **2003**, *20*, 258–263. [CrossRef] [PubMed]
49. Hildebrandt, M.; Schulz, R.; Hoppe, M.; May, T.; Ebner, A. Postoperative routine EEG correlates with long-term seizure outcome after epilepsy surgery. *Seizure* **2005**, *14*, 446–451. [CrossRef] [PubMed]
50. Jeong, S.W.; Lee, S.K.; Hong, K.S.; Kim, K.K.; Chung, C.K.; Kim, H. Prognostic factors for the surgery for mesial temporal lobe epilepsy: Longitudinal analysis. *Epilepsia* **2005**, *46*, 1273–1279. [CrossRef] [PubMed]
51. Jeha, L.E.; Najm, I.M.; Bingaman, W.E.; Khandwala, F.; Widdess-Walsh, P.; Morris, H.H.; Dinner, D.S.; Nair, D.; Foldvary-Schaeffer, N.; Prayson, R.A.; et al. Predictors of outcome after temporal lobectomy for the treatment of intractable epilepsy. *Neurology* **2006**, *66*, 1938–1940. [CrossRef] [PubMed]
52. Velasco, A.L.; Velasco, M.; Velasco, F.; Menes, D.; Gordon, F.; Rocha, L.; Briones, M.; Marquez, I. Subacute and chronic electrical stimulation of the hippocampus on intractable temporal lobe seizures: Preliminary report. *Arch. Med. Res.* **2000**, *31*, 316–328. [CrossRef]
53. Radhakrishnan, A.; Abraham, M.; Vilanilam, G.; Menon, R.; Menon, D.; Kumar, H.; Cherian, A.; Radhakrishnan, N.; Kesavadas, C.; Thomas, B.; et al. Surgery for "Long-term epilepsy associated tumors (LEATs)": Seizure outcome and its predictors. *Clin. Neurol. Neurosurg.* **2016**, *141*, 98–105. [CrossRef] [PubMed]
54. Godoy, J.; Luders, H.; Dinner, D.S.; Morris, H.H.; Wyllie, E.; Murphy, D. Significance of sharp waves in routine EEGs after epilepsy surgery. *Epilepsia* **1992**, *33*, 285–288. [CrossRef] [PubMed]
55. Mintzer, S.; Nasreddine, W.; Passaro, E.; Beydoun, A. Predictive value of early EEG after epilepsy surgery. *J. Clin. Neurophysiol.* **2005**, *22*, 410–414. [PubMed]
56. Kelemen, A.; Rasonyi, G.; Szucs, A.; Fabo, D.; Halasz, P. Predictive factors for the results of surgical treatment in temporal lobe epilepsy. *Ideggyogy Sz* **2006**, *59*, 353–359. [PubMed]
57. Thorbecke, R.; May, T.W.; Koch-Stoecker, S.; Ebner, A.; Bien, C.G.; Specht, U. Effects of an inpatient rehabilitation program after temporal lobe epilepsy surgery and other factors on employment 2 years after epilepsy surgery. *Epilepsia* **2014**, *55*, 725–733. [CrossRef] [PubMed]
58. Vickrey, B.G.; Hays, R.D.; Rausch, R.; Sutherling, W.W.; Engel, J., Jr.; Brook, R.H. Quality of life of epilepsy surgery patients as compared with outpatients with hypertension, diabetes, heart disease, and/or depressive symptoms. *Epilepsia* **1994**, *35*, 597–607. [CrossRef] [PubMed]
59. Chovaz, C.J.; McLachlan, R.S.; Derry, P.A.; Cummings, A.L. Psychosocial function following temporal lobectomy: Influence of seizure control and learned helplessness. *Seizure* **1994**, *3*, 171–176. [CrossRef]
60. Augustine, E.A.; Novelly, R.A.; Mattson, R.H.; Glaser, G.H.; Williamson, P.D.; Spencer, D.D.; Spencer, S.S. Occupational adjustment following neurosurgical treatment of epilepsy. *Ann. Neurol.* **1984**, *15*, 68–72. [CrossRef] [PubMed]
61. Hamiwka, L.; Macrodimitris, S.; Tellez-Zenteno, J.F.; Metcalfe, A.; Wiebe, S.; Kwon, C.S.; Jette, N. Social outcomes after temporal or extratemporal epilepsy surgery: A systematic review. *Epilepsia* **2011**, *52*, 870–879. [CrossRef] [PubMed]
62. Puka, K.; Smith, M.L. Where are they now? Psychosocial, educational, and vocational outcomes after epilepsy surgery in childhood. *Epilepsia* **2016**, *57*, 574–581. [CrossRef] [PubMed]
63. Sperling, M.R.; O'Connor, M.J.; Saykin, A.J.; Plummer, C. Temporal lobectomy for refractory epilepsy. *JAMA* **1996**, *276*, 470–475. [CrossRef] [PubMed]

64. Guldvog, B. Patient satisfaction and epilepsy surgery. *Epilepsia* **1994**, *35*, 579–584. [CrossRef] [PubMed]
65. Reeves, A.L.; So, E.L.; Evans, R.W.; Cascino, G.D.; Sharbrough, F.W.; O'Brien, P.C.; Trenerry, M.R. Factors associated with work outcome after anterior temporal lobectomy for intractable epilepsy. *Epilepsia* **1997**, *38*, 689–695. [CrossRef] [PubMed]
66. Lindsay, J.; Ounsted, C.; Richards, P. Long-term outcome in children with temporal lobe seizures. V: Indications and contra-indications for neurosurgery. *Dev. Med. Child Neurol.* **1984**, *26*, 25–32. [CrossRef] [PubMed]
67. Malkki, H. Epilepsy: Beyond seizure control—Vocational outcomes after paediatric epilepsy surgery. *Nat. Rev. Neurol.* **2016**, *12*, 128. [CrossRef] [PubMed]
68. Jones, J.E.; Blocher, J.B.; Jackson, D.C. Life outcomes of anterior temporal lobectomy: Serial long-term follow-up evaluations. *Neurosurgery* **2013**, *73*, 1018–1025. [CrossRef] [PubMed]
69. Vickrey, B.G.; Berg, A.T.; Sperling, M.R.; Shinnar, S.; Langfitt, J.T.; Bazil, C.W.; Walczak, T.S.; Pacia, S.; Kim, S.; Spencer, S.S. Relationships between seizure severity and health-related quality of life in refractory localization-related epilepsy. *Epilepsia* **2000**, *41*, 760–764. [CrossRef] [PubMed]
70. Reid, K.; Herbert, A.; Baker, G.A. Epilepsy surgery: Patient-perceived long-term costs and benefits. *Epilepsy Behav.* **2004**, *5*, 81–87. [CrossRef] [PubMed]
71. Asztely, F.; Ekstedt, G.; Rydenhag, B.; Malmgren, K. Long term follow-up of the first 70 operated adults in the Goteborg Epilepsy Surgery Series with respect to seizures, psychosocial outcome and use of antiepileptic drugs. *J. Neurol. Neurosurg. Psychiatry* **2007**, *78*, 605–609. [CrossRef] [PubMed]
72. Vickrey, B.G.; Hays, R.D.; Engel, J., Jr.; Spritzer, K.; Rogers, W.H.; Rausch, R.; Graber, J.; Brook, R.H. Outcome assessment for epilepsy surgery: The impact of measuring health-related quality of life. *Ann. Neurol.* **1995**, *37*, 158–166. [CrossRef] [PubMed]
73. Jones, J.E.; Berven, N.L.; Ramirez, L.; Woodard, A.; Hermann, B.P. Long-term psychosocial outcomes of anterior temporal lobectomy. *Epilepsia* **2002**, *43*, 896–903. [CrossRef] [PubMed]
74. Alonso, N.B.; Mazetto, L.; de Araujo Filho, G.M.; Vidal-Dourado, M.; Yacubian, E.M.; Centeno, R.S. Psychosocial factors associated with in postsurgical prognosis of temporal lobe epilepsy related to hippocampal sclerosis. *Epilepsy Behav.* **2015**, *53*, 66–72. [CrossRef] [PubMed]
75. Andersson-Roswall, L.; Engman, E.; Samuelsson, H.; Malmgren, K. Psychosocial status 10 years after temporal lobe resection for epilepsy, a longitudinal controlled study. *Epilepsy Behav.* **2013**, *28*, 127–131. [CrossRef] [PubMed]
76. Taylor, D.C.; McMackin, D.; Staunton, H.; Delanty, N.; Phillips, J. Patients' aims for epilepsy surgery: Desires beyond seizure freedom. *Epilepsia* **2001**, *42*, 629–633. [CrossRef] [PubMed]
77. Baca, C.B.; Cheng, E.M.; Spencer, S.S.; Vassar, S.; Vickrey, B.G. Racial differences in patient expectations prior to resective epilepsy surgery. *Epilepsy Behav.* **2009**, *15*, 452–455. [CrossRef] [PubMed]
78. Ozanne, A.; Graneheim, U.H.; Ekstedt, G.; Malmgren, K. Patients' expectations and experiences of epilepsy surgery—A population-based long-term qualitative study. *Epilepsia* **2016**, *57*, 605–611. [CrossRef] [PubMed]
79. Wilson, S.J.; Saling, M.M.; Kincade, P.; Bladin, P.F. Patient expectations of temporal lobe surgery. *Epilepsia* **1998**, *39*, 167–174. [CrossRef] [PubMed]

behavioral sciences

MDPI

Article

A Cohort Study Comparing Women with Autism Spectrum Disorder with and without Generalized Joint Hypermobility

Emily L. Casanova [1,2,*], Julia L. Sharp [3], Stephen M. Edelson [4], Desmond P. Kelly [2] and Manuel F. Casanova [1,2]

[1]   Department of Biomedical Sciences, University of South Carolina School of Medicine Greenville, Greenville, SC 29605, USA; mcasanova@ghs.org
[2]   Department of Pediatrics, Greenville Health System Children's Hospital, Greenville, SC 29605, USA; dkelly@ghs.org
[3]   Department of Statistics, Colorado State University, Fort Collins, CO 80523, USA; julia.sharp@colostate.edu
[4]   Autism Research Institute (ARI), San Diego, CA 92116, USA; director@autism.com
*   Correspondence: casanove@greenvillemed.sc.edu

Received: 23 November 2017; Accepted: 15 March 2018; Published: 17 March 2018

Abstract: Reports suggest comorbidity between autism spectrum disorder (ASD) and the connective tissue disorder, Ehlers-Danlos syndrome (EDS). People with EDS and the broader spectrum of Generalized Joint Hypermobility (GJH) often present with immune- and endocrine-mediated conditions. Meanwhile, immune/endocrine dysregulation is a popular theme in autism research. We surveyed a group of ASD women with/without GJH to determine differences in immune/endocrine exophenotypes. ASD women 25 years or older were invited to participate in an online survey. Respondents completed a questionnaire concerning diagnoses, immune/endocrine symptom history, experiences with pain, and seizure history. ASD women with GJH (ASD/GJH) reported more immune- and endocrine-mediated conditions than their non-GJH counterparts ($p = 0.001$). Autoimmune conditions were especially prominent in the ASD/GJH group ($p = 0.027$). Presence of immune-mediated symptoms often co-occurred with one another ($p < 0.001$–0.020), as did endocrine-mediated symptoms ($p < 0.001$–0.045), irrespective of the group. Finally, the numbers of immune- and endocrine-mediated symptoms shared a strong inter-relationship ($p < 0.001$), suggesting potential system crosstalk. While our results cannot estimate comorbidity, they reinforce concepts of an etiological relationship between ASD and GJH. Meanwhile, women with ASD/GJH have complex immune/endocrine exophenotypes compared to their non-GJH counterparts. Further, we discuss how connective tissue regulates the immune system and how the immune/endocrine systems in turn may modulate collagen synthesis, potentially leading to higher rates of GJH in this subpopulation.

Keywords: connective tissue diseases; autoimmunity; mast cells; immunity; humoral; endocrine system diseases; neurodevelopmental disorders

1. Introduction

Ehlers-Danlos syndrome (EDS) is a group of phenotypically-related disorders subtyped according to variations in underlying genetic pathology, primary symptom severity, and secondary symptom associations. All of these conditions are typified by deficits in collagen production and maintenance, leading to structural changes within the connective tissues of the body. These changes are most evident within the joints and skin, although many other systems can be affected.

Generalized joint hypermobility (GJH) is a major feature of Hypermobile EDS (hEDS) and other connective tissue disorders. In addition, GJH can either be benign or associated with significant musculoskeletal impairment; the latter of which is often affected by an individual's age, leading to changes in diagnosis over time. According to newer nosology [1], when GJH occurs in conjunction with significant impairment and other criteria for hEDS are not met, it is diagnosed as "Generalized Hypermobility Spectrum Disorder (G-HSD)." Studies have shown that hEDS and GJH often co-segregate within families, indicating linked etiologies in some cases [reviewed in 1].

Neuropsychiatric manifestations are common secondary symptoms in EDS/GJH. In particular, anxiety and mood disorders are prominent and probably the best studied to date [2,3]. However, a thorough review of the literature by Baeza-Velasco et al. [4] suggests significant links between EDS and autism, as well as other neurodevelopmental and psychiatric conditions such as attention-deficit hyperactivity disorder (ADHD), schizophrenia, eating disorders, personality disorders, and even substance abuse. Interestingly, work by Shetreat et al. [5] and Eccles et al. [6] indicates that joint hypermobility is significantly more common in children and adults with autism than age- and gender-matched controls, suggesting etiological links between some cases of autism and connective tissue disorders.

Immune & Endocrine Dysregulation in ASD & GJH

Immune and neuroendocrine crosstalk is a well-established phenomenon. These systems are linked via two primary pathways through which that crosstalk is achieved: (1) the sympathoadrenal system and (2) the hypothalamo-pituitary-adrenal axis [7]. The immune system can also have a direct effect on oogenesis through the presence of innate and adaptive immune cells located within the ovarian germ cell pool, which release morphoregulatory signals that stimulate or suppress ovulation [8].

Immune dysregulation has been a popular area of study in autism research, whose foci center around topics of maternal immune activation (MIA), prevalence of autoimmunity, and other aspects of general immune dysfunction [9]. In regards to the latter, Careaga et al. [10] have identified two non-overlapping Th1- and Th2-skewed endophenotypes that are especially prominent in children with ASD.

Hormonal exophenotypes, in contrast, have been less well-studied in ASD. One study by Ingudomnokul et al. [11] found that high-functioning women on the autism spectrum and their mothers reported high rates of endocrine disorders. However, most endocrine research to date has focused on maternal disorders with an emphasis on etiological risk factors, such as diabetes, hirsutism, and polycystic ovary syndrome (PCOS) [12–14].

High rates of immune- and endocrine-mediated disorders have also been reported in EDS, though they are currently viewed as secondary symptoms to what are traditionally seen as "collagen disorders" [15–17]. While it has previously been difficult to explain links between immune and collagen dysfunction, research into the connective tissue disorders, Marfan and Loeys-Dietz Syndromes, which share features of overlap with EDS, may help to guide future EDS research.

In order to study the frequency and relationship of immune and endocrine exophenotypes in adult women with ASD, with or without GJH, we have utilized self-reports covering a range of clinical symptoms, including features of chronic allergies, autoimmunity, irritable bowel syndrome (IBS)/gastrointestinal (GI) dysfunction, and menstrual irregularities.

2. Methods

2.1. Study Population

The vast majority (94%) of respondents were affected persons themselves, rather than family members responding for adult wards. As such, it is assumed that the majority of our study population was composed of women with an IQ > 70 due to their abilities to answer a series of complex questions about general health.

Our study group was composed of two English-speaking subpopulations: (1) women 25 years and older with a diagnosis of ASD (referred to here as simply "ASD") (N = 85); and (2) women 25 years or older with dual ASD and EDS, G-HSD, or Joint Hypermobility Syndrome (JHS) diagnoses (referred to here as "ASD/GJH") (*N* = 20) (Figure 1). Individuals who were male, were under the age of 25, or did not have a diagnosis of ASD were excluded. In the ASD group, further exclusionary criteria were applied: (a) An individual's responses were removed from the data pool if she suspected the presence of GJH but was currently undiagnosed, and/or (b) reported double-jointedness across two or more types of joints [18]. The majority of women reporting EDS diagnosis had hEDS, although a small minority reported diagnosis of Classical EDS. (See Table 1 for descriptions of terms and definitions.)

Our groups were sex-matched and did not differ significantly by age (*t* = −0.327, *df* = 28.451, *p* = 0.7459). Full data are presented in Supplementary File 2, Tables S1–S7. All data were complete, with the exception of two respondents' answers on the topic of "Other Chronic Pain".

Due to the biased manner in which respondents were recruited [i.e., specifically targeting both ASD and ASD/GJH subgroups via respective web fora (see Section 2.2 under Methods)], we are unable to estimate the prevalence of GJH in the female ASD population. However, this method allowed us to collect a larger pool of ASD/GJH respondents, which might otherwise be underrepresented. In doing so, we are able to study group differences more easily.

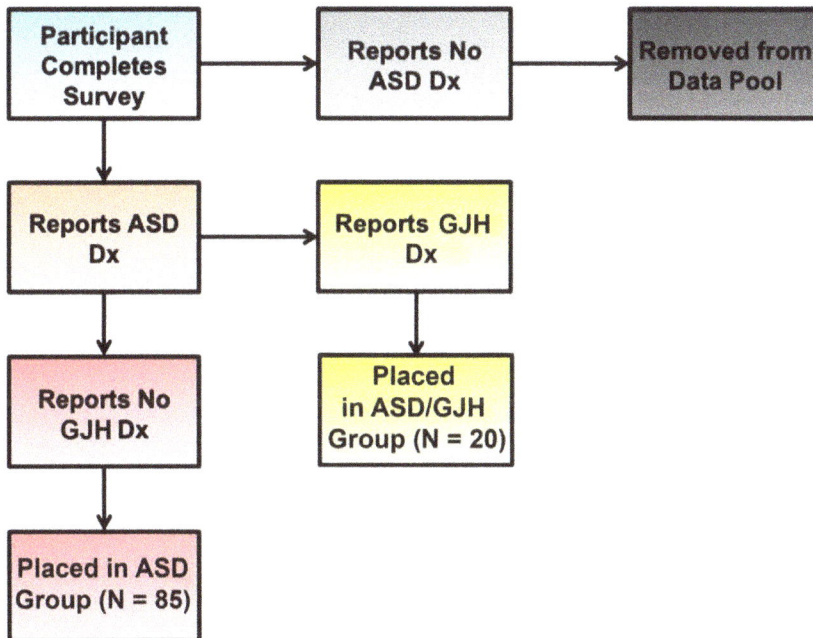

Figure 1. Flow chart illustrating group allocation according to reported diagnoses.

Table 1. Terms and diagnoses related to the connective tissue disorders discussed in the manuscript.

Generalized Joint Hypermobility-Related Diagnoses	Description
Hypermobile Ehlers-Danlos Syndrome (hEDS)—Formerly known as EDS, Hypermobile Type, or EDS Type III.	• Generalized joint hypermobility • Musculoskeletal involvement (arthralgia, instability) • Involvement of other organ systems (skin, Marfanoid features, etc.) • No consistently associated gene mutations

Table 1. *Cont.*

Generalized Joint Hypermobility-Related Diagnoses	Description
Classical Ehlers-Danlos Syndrome (cEDS)—Also known as EDS Type I.	• Skin hyperextensibility and atrophic scarring • Generalized joint hypermobility • Minor features: e.g., easy bruising, skin fragility, hernias, etc. • Associated gene mutations: *COL1A1*, *COL5A1*, and *COL5A2*
Generalized Hypermobility Spectrum Disorder (G-HSD) - Formerly known as "non-benign" JHS.	• Generalized joint hypermobility • Musculoskeletal involvement (arthralgia, instability) • Other minor criteria associated with hEDS may be present but to a comparatively lesser extent
* Joint Hypermobility Syndrome (JHS)—Divided into "benign" and "non-benign" forms. Diagnosis now in disuse as of 2017.	• Generalized joint hypermobility • Optional: musculoskeletal involvement (arthralgia, instability)
Hypermobility Spectrum Disorders (HSD)	Composed of: • G-HSD (formerly known as "non-benign" JHS) • Peripheral HSD (P-HSD) • Localized HSD (L-HSD) • Historical HSD (H-HSD)
Asymptomatic Joint Hypermobility	• Asymptomatic Generalized Joint Hypermobility (A-GJH) (formally known as "benign" JHS) • Asymptomatic Peripheral Joint Hypermobility (A-PJH) • Asymptomatic Localized Joint Hypermobility (A-LJH)
Marfan Syndrome (MFS)	• Aortic root dilation • Ectopia lentis (dislocated lenses of the eye) • Minor features: Marfan habitus, generalized joint hypermobility • Associated gene mutations: *FBN1*
Loeys-Dietz Syndrome (LDS)	• Enlargement of the aorta • Aneurysms • Hypertelorism • Bifid uvula or cleft palate • Minor features: Marfanoid habitus, immune disorders (allergy, asthma, rhinitis, eczema) • Associated gene mutations: *TGFBR1*, *TGFBR2*, *SMAD3*, *TGFB2*, and *TGFB3*

* indicates terminology that is no longer in use as of the recent nosological changes enacted [ ].

2.2. Survey

This study was approved by the Institutional Review Board (IRB) of the Greenville Health System (GHS) (ID: Pro00061122). The survey utilized in this study was designed by our research group based in part on previous informal survey studies performed by the Autism Research Institute (ARI). These questions were further adapted and expanded according to an additional literature search of relevant clinical symptomology for our topics of interest. (See Supplementary File 1 for full survey.)

The survey was built on and hosted by the website, SurveyGizmo.com, and was advertised via the ARI newsletter; the ASD forum, Wrong Planet; and a variety of FaceBook ASD- and

EDS-specific webcommunities, such as the Autism Women's Network (AWN), the Autism Spectrum Women's Group, AutismTalk, the Ehlers-Danlos Support Group, and Ehlers-Danlos Worldwide. The administrative teams of all participating web communities were informed that the survey was IRB-approved and were given access to the survey and, when requested, a copy of the IRB protocol prior to approval. Following administrator approval, either ELC posted the survey announcement or administrators posted it themselves. The survey weblink (www.autismwomensstudy.com) led potential respondents to a description of the purpose and expectations of the study, potential risks and benefits, investigator contact information, and a waiver of consent. The survey was open and participants were actively recruited for approximately three months.

Survey questions focused on topics concerning ASD and EDS/GJH diagnoses; symptoms involving the immune and endocrine systems; chronic pain; GI dysfunction such as IBS; seizures; and limited aspects of medication history (hormone treatment, antiseizure medications). Additional topics were covered but were not used for the current study.

Questions on immune symptomology included items concerning hospitalization history; respiratory disorders like asthma, allergies, sinusitis/rhinitis, and reactions to medications or environmental chemicals; and autoimmunity [19–21]. Hormone-mediated symptoms included items such as chronic irregularities in menstruation and associated pain syndromes; PCOS; and other clinical symptoms indicative of the metabolic syndrome, such as type 2 diabetes/insulin resistance, hypertension, and high cholesterol [11,22]. The data of respondents who agreed to participate but failed to complete the survey were discarded.

2.3. Statistical Analyses

When assessing group differences for quantitative variables (e.g., age, immune- and endocrine-mediated symptoms), Welch two-sample t-tests were conducted unless the distributions were heavily skewed, in which case the Wilcoxon rank sum test with continuity correction was used. Two sample tests of proportions were used to compare groups for binary categorical data (e.g., presence/absence of diabetes, infertility, etc.). Fisher's Exact Test was used in cases of small sample sizes. Where appropriate, a false discovery rate adjustment was used to account for multiple comparisons. A significance level of 0.05 was used for all analyses.

3. Results

3.1. Immune-Mediated Disorders

Although women with ASD and ASD/GJH did not differ in the presence of one or more immune-mediated symptoms ($\chi^2 = 1.162$, $p = 0.281$), ASD/GJH women were, however, more likely to report multiple symptoms ($t = -3.860$, $df = 30.981$, $p = 0.001$), an effect that differed by age ($W = 534.5–563$, $p = 0.009–0.017$) (Figure 2A). Women with ASD and ASD/GJH also reported similar proportions of specific immune exophenotypes ($\chi^2 = 0.788–4.744$, $p = 0.137–0.375$), with an overall trend towards higher proportions in the ASD/GJH group. However, one exception concerned autoimmune disorders: while 13% of the ASD group had an autoimmune disorder, 45% of women with ASD/GJH reported the same ($\chi^2 = 8.813$, $p = 0.027$) (Table 2).

The presence of most immune exophenotypes exhibited a significant association with one another, suggesting that a similar etiological background underlies many of these symptoms. Allergies, rhinitis, sinusitis, asthma, ear infections, reaction to medications, and reaction to environmental chemicals all seemed to share a strong interrelationship ($p < 0.001–0.020$). Meanwhile, autoimmunity was significantly associated with ear infections ($CI = 1.509–18.898$, $p = 0.014$) and showed a trend towards significance with asthma ($CI = 1.003–9.582$, $p = 0.059$) (Figure 2B).

Table 2. Reported rates of various immune-related symptoms according to group, as well as estimated general prevalence rates.

Immune Symptomology	ASD (N = 85)	ASD/EDS (N = 20)	General Prevalence
Allergies	45%	60%	30% in adults [23]
Asthma	33%	60%	8.4% in general population [24]
Autoimmunity	13%	45% *	7.6–9.4% in general population [25]
			83% with ≥1 incidents between 0–3
Chronic Ear Infections	40%	65%	years of age [26] 11% with ≥1 incidents of all ages [27]
Chronic Rhinitis	38%	60%	8% in adults [28]
Chronic Sinusitis	46%	60%	8% in adults [28]
Severe Reaction to Medications	35%	65%	10–15% of hospitalized patients [29]
Severe Reaction to Environmental Chemicals	39%	65%	13–16% in adults [30]

* indicates a significant difference between groups.

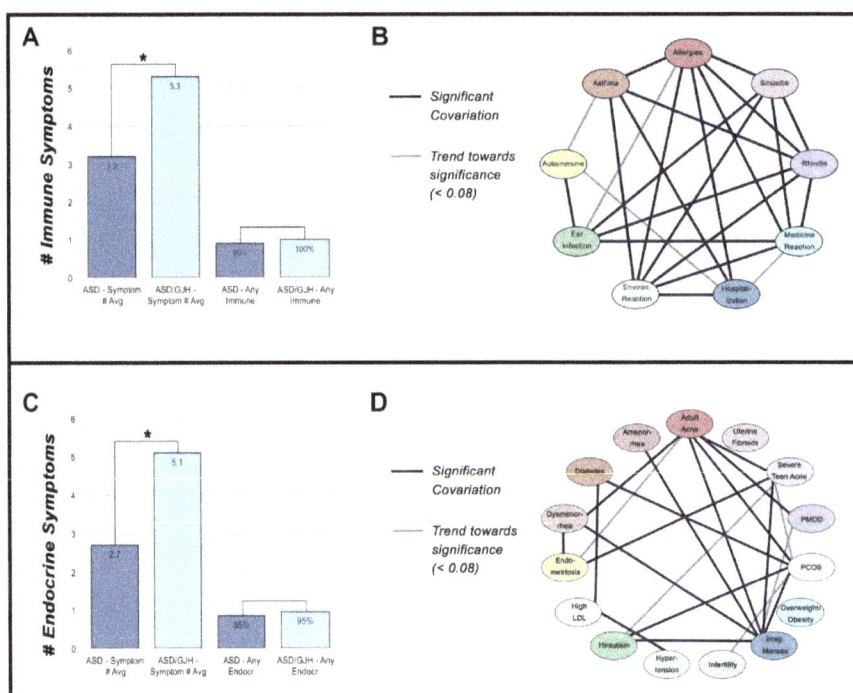

Figure 2. (**A**) Number of immune-mediated symptoms across ASD and ASD/GJH groups. 'Any immune' = 1 or more immune symptoms. (**B**) Network of immune-mediated symptoms. (**C**) Number of endocrine-mediates symptoms across ASD and ASD/GJH groups. 'Any endocr.' = 1 or more endocrine symptoms. (**D**) Network of endocrine-mediated symptoms.

Interestingly, the proportions of IBS/GI dysfunction did not differ significantly between groups, though the ASD/GJH group reported modestly higher rates (χ^2 = 0.648, p = 0.946). In spite of IBS' links with immunity, it did not share a significant relationship with immune exophenotypes in general (CI = 0.282–7.60, p = 0.717), although our study may have been too underpowered to glean an effect [31]. Yet in spite of modest numbers, IBS/GI dysfunction was significantly linked with hormonal exophenotypes: individuals with IBS/GI dysfunction had more hormone-mediated symptoms on average than those without (W = 811, p = 0.002).

In spite of the previously reported relationship between hormones and seizure propensity, the presence of complex hormonal exophenotypes was not associated with epilepsy in our cohort (W = 274, *unadj.* p = 0.3742) [32]. However, despite the small number of women reporting epilepsy (N = 7), epilepsy shared a modest positive relationship with the number of immune-mediated disorders irrespective of the group (W = 166, *unadj.* p = 0.022). While the average number of immune-mediated disorders reported across the entire cohort was approximately 3.6 (SD = 2.45), women with epilepsy averaged approximately 5.7 (SD = 2.43).

Finally, joint pain was reported in all cases of ASD/GJH compared to 29% in the ASD group (χ^2 = 30.122, p < 0.001). Meanwhile, differences in joint pain were not accounted for by age (W = 559.5, p = 0.101) or obesity (χ^2 = 0, p = 1.000). Other types of chronic pain were also reported more often in ASD/GJH (75% vs. 31%) (χ^2 = 11.072, p < 0.001), including conditions such as fibromyalgia [33].

3.2. Hormone Disorders

Though the ASD/GJH and ASD groups did not differ in the presence of one or more hormone-mediated disorders (χ^2 = 0.728, p = 0.394), ASD/GJH women reported significantly more symptoms than their non-GJH counterparts (W = 434, p = 0.001) (Figure 2C). On a symptom-by-symptom basis, ASD/GJH women reported higher rates of endometriosis (χ^2 = 9.265, p = 0.018), dysmenorrhea (χ^2 = 19.599, p < 0.001), and severe teen acne (χ^2 = 7.817, p = 0.026) (Table 3). Dysmenorrhea, in particular, was reported three times more often (85% vs. 28%) in ASD/GJH compared to ASD (χ^2 = 19.60, p < 0.001), a frequency similar to that reported in previous EDS research (see Table 3). In its extreme form, dysmenorrhea is typically associated with endometriosis and both share links with immune dysfunction in the general population [34,35].

Table 3. Reported rates of various endocrine-related symptoms according to group, as well as estimated general prevalence rates.

Endocrine Symptomology	ASD (N = 85)	ASD/EDS (N = 20)	General Prevalence
Adult Acne	21%	35%	35% in women ages 30–39 [36]
Amenorrhea	39%	45%	4.6% in women ages 15–44 [37]
Diabetes/Insulin Resistance	6%	10%	7.9% in adults [38]
Dysmenorrhea	28%	85% *	2–29% in adult women [39]
Endometriosis	5%	30% *	4% in women [40]
High LDL Cholesterol	14%	30%	28% in adults [41]
Hirsutism	19%	30%	10% in adult women [42]
Hypertension	14%	20%	29.1% in adults [43]
Infertility	8%	15%	6% [44]
Irregular Menstruation	27%	55%	18.2% in adult women [45]
Overweight/Obesity	36%	45%	70.7% aged 20+ years [46]
Polycystic Ovary Syndrome (PCOS)	8%	25%	7.3% in adult women [47]
Premenstrual Dysphoric Disorder (PMDD)	21%	30%	3–8% of premenopausal women [48]
Severe Teen Acne	14%	45% *	12.1% in males and females aged 17 [49]
Uterine Fibroids	9%	5%	4.5–9.8% in adult women aged 40–49 [50]

* indicates a significant difference between groups.

We found no significant interaction between group and birth control/hormone treatment in relation to the average number of hormone-mediated symptoms, indicating that such treatment is an unlikely group confound in this study. There was, however, a significant relationship between the number of hormone-mediated symptoms reported and whether an individual was receiving some form of hormonal treatment ($t(101) = 2.75$, $p = 0.004$). While we cannot rule out a potential confound, instead, we conclude that this is likely a reflection of the severity of endocrine disorders in our cohort and their prescribed treatments [51].

Like immune symptoms, individual hormone-mediated symptoms were often associated with one another (Figure 2D). PCOS shared links with other symptoms, including diabetes, adult acne, irregular menses, and hirsutism, all of which are either diagnostic of or commonly reported in PCOS ($p = 0.005$–0.045). In contrast, infertility, overweight/obesity, amenorrhea, hypertension, and high low-density lipoprotein (LDL) cholesterol did not associate with PCOS in our groups ($p = 0.073$–0.809) [52]. There was, however, a trend towards significance between PCOS and infertility, suggesting our data may have been underpowered, requiring a larger pool of respondents in the future (OR 95% $CI = 1.20$–37.800, $p = 0.073$). Endometriosis and dysmenorrhea were also associated with one another ($CI = 2.229$–785.323, $p = 0.013$).

3.3. The Relationship between Immune- & Hormone-Mediated Symptoms

There was no significant association between the general presence of immune- and endocrine-mediated disorders across our cohort (OR 95% $CI = 0.538$, 21.119, $p = 0.098$). However, the number of immune-mediated symptoms per individual greatly predicted the number of hormone-mediated symptoms (*Spearman's rho* $= 0.35$, $p < 0.001$). This suggests that the complexity and severity of immune- and endocrine-mediated disorders share a strong positive relationship with one another in autism and potentially within the general population, e.g., [53].

4. Discussion

The present study attempts to address phenotypic differences between ASD women with and without GJH. This research supports a growing body of literature indicating that immune-mediated disorders are a common comorbid feature in hEDS and GJH. In addition, we have also shown that this dysfunction may be paired with endocrine dysregulation, leading to complex immune and hormonal exophenotypes, such as autoimmune disorders, allergic rhinitis, asthma, endometriosis, and dysmenorrhea. While we have not addressed autism and GJH comorbidity rates in this study, their co-occurrence in the adult ASD female population suggests links between the dysfunction of connective tissue and the immune and endocrine systems in this subpopulation.

As discussed, the immune system has been a popular area of investigation in autism research. However, reports of clinical manifestations in the child population seem to vary [54–57]. Some clinical manifestations arise during or progress in severity with the advent of puberty, highlighting the role the endocrine system plays in immune function, e.g., [58]. In addition, women are more frequent targets of such dysfunction, suggesting that studying immune dysregulation in prepubertal individuals with autism, while also ignoring gender confounds, dramatically underrepresents the frequency of clinical symptoms in the autism population [19]. For these reasons, we limited our study population to women aged 25 years or older on the autism spectrum.

4.1. Immune-Mediated Disorders in Association with Connective Tissue Disorders

Loeys-Dietz Syndrome (LDS) is a connective tissue disorder caused by mutations directly targeting the TGF-β pathway and is characterized primarily by enlargement of the aorta. People with LDS have high rates of immune-mediated disorders such as respiratory and food allergies and occasionally present with Hyper-IgE Syndrome, a type of primary immunodeficiency [59]. In addition, they also share many of the same dysmorphic features as those seen in the connective tissue disorder, Marfan Syndrome (MFS) [60].

Although MFS is associated with mutations in the *Fibrillin-1* (*FBN1*) gene whose protein product is a component of the extracellular matrix (ECM), *FBN1* mutations lead to marked TGF-β dysregulation [61–63]. Fibrillin appears to control the activity of TGF-β by acting as a structural platform for the Latent TGF-β Binding Protein (LTBP) that sequesters and inactivates TGF-β, acting as a reserve pool for rapid injury response [64]. Given its role as a foundational morphogen, it is believed this overlap in TGF-β pathway dysregulation leads to the overlapping features of MFS and LDS [65].

Like LDS, some individuals with hEDS present with a Marfanoid (Marfan-like) habitus [66]. However, unlike MFS that results from dysfunctional fibrillin, EDS is typically linked with dysfunction of the ECM protein, collagen. Marfan and Marfanoid features in all three of these disorders suggest considerable overlap and interaction between the ECM and the TGF-β pathway. In addition, TGF-β serves as a link between the ECM and immune system disruption as it is a key immunomodulator, implicated not only within the joints in these connective tissue disorders, but also in other organ systems such as the lungs [65]. Interestingly, several studies have consistently found lower TGF-β1 levels in autism, which according to Ashwood et al. [67], may help explain some of the immune dysregulation in the condition [67,68]. For these reasons, the TGF-β pathway and upstream networks may be prime areas of study for future work into the overlapping etiologies of both connective tissue disorders and autism.

4.2. The Effects of Estrogen on Collagen Production & the Immune System

Similar to certain immune disorders like autoimmunity, GJH and hEDS preferentially target women for reasons not well understood [69]. One possibility may stem from sex differences in muscle mass, in which stronger muscles help to counteract joint laxity and ensuant pain [16]. For this reason, one of the foci of physical therapy in the treatment of GJH/hEDS centers around improved muscle strength surrounding problem joints [70]. However, female-specific effects may result not only from low testosterone levels, but also estrogen metabolites that either suppress collagen production directly, particularly within the skin, or result in a more rapid turnover of collagen within tendons and ligaments [71–73].

Estrogen is also a major immunomodulator. It is capable of driving activation of the Th2 branch of the immune system, boosting humoral immunity and the ability of the body to target parasites and other extracellular infections. Estrogen also stimulates mast cell degranulation, prompting a release of chemicals such as histamines, TNF-α, various amines, chymase, and tryptase [74,75]. Mast cell activation, in turn, may drive both Th1/Th2 immune responses depending on the invading pathogen, the target tissue, and other variable factors [76].

Interestingly, estrogen also increases the synthesis of TGF-β within numerous cell types, the latter of which is itself a key morphogen and immunodulator. In addition, estrogen further interacts with the TGF-β pathway by forming a complex with Smad 3/4, redirecting TGF-β target genes. Finally, TGF-β and estrogen are able to interact at the level of various Ras complexes, by which TGF-β enhances estrogenic action [77]. All of these data together suggest significant interaction of estrogen with various networks implicated in connective tissue disorders and their secondary symptoms.

4.3. Autism & Generalized Joint Hypermobility

Results of this study indicate that the ASD/GJH phenotype in women is characterized not only by classic symptoms of EDS/G-HSD such as generalized hypermobility and chronic pain, but that immune and endocrine system involvements may be extensive. In addition, phenotypic expression of this immune disorder is mediated by the endocrine system and the ongoing presentation of symptoms throughout life are guided by immune-endocrine crosstalk.

In support of this, all 20 ASD/GJH women in our study group reported ≥ 2 immune-mediated symptoms, with an average reporting of 5.3 symptoms per person compared to 3.2 in the ASD group. Likewise, 90% of ASD/GJH women reported ≥ 2 hormone-mediated symptoms, with an average of 5.1, compared to 2.7 in ASD. Therefore, the vast majority of ASD/GJH women in this study reported multiple immune- and endocrine-mediated symptoms, the extent of which appears to vary with one another.

Mast Cell Activation Syndrome (MCAS), a newly recognized diagnostic entity with growing clinical significance, may be relevant to immune exophenotypes reported by our participants [15]. While the traditional slew of MCAS impairments include analphylaxis, syncope, flushing, urticaria, and GI distress (e.g., diarrhea, nausea, vomiting), continued study of this condition reveals a broader spectrum of physical ailments relative to the locations of mast cells involved, the extent of stimulation, and the specific mediators released.

Although MCAS can mimic many localized diseases, its defining feature is chronic mast cell activation across two or more organ systems, which is reminiscent of the complex combination of respiratory, connective tissue, and GI symptoms reported by some of our participants [78,79]. Interestingly, MCAS is also a common comorbid feature of EDS and postural orthostatic syndrome (POTS), reinforcing this emerging pattern [15,80]. Current prevalence rates of this newly recognized entity (14–17%) also suggest it is far more common in the general population than originally believed [78].

While GJH can occur without complications, many cases involve extensive inflammation at the affected joints, suggesting a potential immune component in the disorder as is seen in TGF-β pathway involvement in LDS and MFS. As Afrin [78] suggests in reference to the MCAS-/hEDS relationship:

> ... *chronic aberrant elaboration of a particular set of mediators (drawn from amongst the mast cell's repertoire of more than 200 such molecular signals) not only [influences] virtually every other system and organ in the body but also [influences] connective tissue development to yield the "hyperextensible" phenotype long associated with EDS Type III [(hypermobile type)]. (p. 138)*

4.4. The Etiology of Autism

While this study cannot address rates of ASD and GJH co-occurrence because of the way in which respondents were recruited, the comorbidity itself reinforces etiological links between autism and connective tissue disorders. Both cytokines and hormones play recognized roles in neurogenesis, neuritogenesis, synaptogenesis, and ongoing plasticity [81–84]. In addition, some researchers have proposed that autoantibodies to brain-specific proteins may also disrupt neurodevelopment, leading to increased autism risk [85]. Finally, endocrine disruption, either via endogenous or exogenous effectors, is likewise a growing area of research into autism's etiology [12,86]. All of these topics highlight the crosstalk between the immune and endocrine systems and strengthen their combined links to ASD.

4.5. Limitations

According to recent changes in nosology, hEDS, the most common of the Ehlers-Danlos Syndromes, lies on a continuum with Hypermobility Spectrum Disorders (HSD), including what was once known as Joint Hypermobility Syndrome (JHS) (see Table 1). Previous studies have shown that hEDS and JHS often co-segregate within families, suggesting that in some cases, JHS/HSD may be a lighter variant of hEDS (reviewed in [1]).

As of last year, the criteria for hEDS have become more stringent, placing greater focus on the additional involvement of tissue systems outside that of the musculoskeletal system, e.g., skin and other organs [1]. It is therefore possible and probable that some individuals in this study who had a previous diagnosis of EDS, Hypermobile Type, no longer reach the cut-off for hEDS and would instead be given a diagnosis of Generalized HSD (G-HSD) were they reassessed.

Due to the nature of online surveys and our inability to reassess participants for appropriate recategorization, it is therefore assumed that the ASD/GJH group in this study contains a mix of individuals who would currently be defined as G-HSD and hEDS. For these reasons, our results may not be fully applicable to hEDS and must therefore be interpreted with caution.

Other limitations of our study concern the reliability of data derived from self-reports, which is vulnerable to reporting bias. In particular, the similarity between rates of clinical presentation in our ASD group and the general population suggests reporting reliability (Tables 2 and 3). Meanwhile, similarly high rates of immune- and endocrine-mediated disorders in our ASD/GJH group compared to the general HSD/EDS population also support the veracity of their reports [17,33,69].

Behav. Sci. **2018**, *8*, 35

A related vulnerability of our data hinges on ASD and GJH diagnostic reliability. While the data is dependent upon self-reports, we did however offer respondents the opportunity to specify whether they were professionally diagnosed or suspected a diagnosis. Those who indicated a suspicion of hEDS or some type of HSD were initially included in the first round of analyses as an additional group of interest. However, their data varied too dramatically from the diagnosed group and were not included in the final analysis. Therefore, while the diagnostics are not standardized in this study, those reporting professional diagnoses of ASD or GJH were assumed to be truthful.

Another limitation concerns small sample sizes, particularly of the ASD/GJH group. Given the rarity of EDS (1:5000) and the infrequency of its overlap with ASD (3%), a sample size of 20 could be considered quite large [87,88]. There are unfortunately no current estimates of G-HSD prevalence under the new nosology; however, our results indicate that we have had ample power for this study.

We selectively surveyed ASD women aged 25 years or older to study specific immune and endocrine exophenotypes. However, we cannot generalize our results to the broader autism spectrum, though previous studies indicate that related endo- and exophenotypes exist in ASD males and individuals under the age of 25. Likewise, we cannot generalize our data to the full EDS and GJH spectrums, though previous research supports our findings [17,79,80,89]. Instead, future research is needed to explore a potential clinical spectrum that spans the sexes and the lifespan to determine to what extent our findings apply to the broader autism spectrum and GJH.

Finally, our results suggest there may be a relationship between epilepsy and immune symptomology, which is supported by the recognized roles that cytokines and other immune factors play in epileptogenesis [90]. However, due to small participant numbers, further investigation is necessary to address this potential and is a topic we will be addressing in future studies.

Supplementary Materials: The following are available online at http://www.mdpi.com/2076-328X/8/3/35/s1, Table S1: immune symptoms: hospitalization history, asthma, ear infections, rhinitis, and sinusitis. 0 = no symptom reported; 1 = symptom reported, Table S2: immune symptoms: allergies, reaction to medications (med react), reaction to environmental chemicals (env react), autoimmunity, and sum of all immune symptoms. 0 = no symptom reported; 1 = symptom reported, Table S3: endocrine symptoms: polycystic ovary syndrome (PCOS), amenorrhea, diabetes 2/insulin resistance (diabetes, endometriosis, and adult acne. 0 = no symptom reported; 1 = symptom reported, Table S4: Endocrine symptoms: infertility, dysmenorrhea, irregular menses, high LDL cholesterol, and hypertension. 0 = no symptom reported; 1 = symptom reported, Table S5: Endocrine symptoms: hirsutism, overweight/obesity, premenstrual dysphoric disorder (PMDD), severe teen acne, uterine fibroids, and sum of all endocrine symptoms. 0 = no symptom reported; 1 = symptom reported, Table S6: Sum of immune symptoms by age range (child, teen, and adult). 0 = no symptom reported; 1 = symptom reported, Table S7: Symptoms of irritable bowel syndrome/gastrointestinal dysmotility (IBS), joint pain, other chronic pain, epilepsy, and birth control/hormone treatment (BC Tx). 0 = no symptom reported; 1 = symptom reported.

Acknowledgments: We would like to give the warmest of thanks to Jeanne Winstead, without whose advice and help in advertising this survey our study would not have been possible. We would also like to thank the many websites, both autism- and EDS-specific, that allowed us to advertise the survey and collect respondents. Thank you to the respondents themselves and their willingness to participate in this study. And finally, thank you to Lawrence Afrin who took time from his busy clinical schedule to read this article and offer valuable critique. This work was supported by the National Institutes of Health [grant number R01 HD-65279].

Author Contributions: E.L.C. conceived of the study. E.L.C. and S.M.E. designed the survey and E.L.C., along with help from the autism and Ehlers-Danlos/Hypermobility Spectrum Disorder communities, worked to advertise the study online. J.L.S. performed the statistical analyses. M.F.C. and D.P.K. provided expertise on autism and were integral in helping to design the overall study. All authors contributed substantially to the drafts and have read and approved the final manuscript.

Conflicts of Interest: E.L.C. reported no biomedical financial interests or potential conflicts of interest. J.L.S. reported no biomedical financial interests or potential conflicts of interest. S.M.E. reported no biomedical financial interests or potential conflicts of interest. D.P.K. reported no biomedical financial interests or potential conflicts of interest. M.F.C. reported no biomedical financial interests or potential conflicts of interest.

References

1.   Castori, M.; Tinkle, B.; Levy, J.; Grahame, R.; Malfait, F.; Hakim, A. A framework for the classification of joint hypermobility and related conditions. *Am. J. Med. Genet. C Semin. Med. Genet.* **2017**, *175*, 148–157. [CrossRef] [PubMed]

2. Sinibaldi, L.; Ursini, G.; Castori, M. Psychopathological manifestations of joint hypermobility syndrome/Ehlers-Danlos syndrome, hypermobility type: The link between connective tissue and psychological distress revised. *Am. J. Med. Genet. C Semin. Med. Genet.* **2015**, *169*, 97–106. [CrossRef] [PubMed]

3. Lumley, M.A.; Jordan, M.; Rubenstein, R.; Tsipouras, P.; Evans, M.I. Psychosocial functioning in the Ehlers-Danlos syndrome. *Am. J. Med. Genet.* **1994**, *53*, 149–152. [CrossRef] [PubMed]

4. Baeza-Velasco, C.; Pailhez, G.; Bulbena, A.; Baghdadli, A. Joint hypermobility and the heritable disorders of connective tissue; Clinical and empirical evidence of links with psychiatry. *Gen. Hosp. Psychiatry* **2015**, *37*, 24–30. [CrossRef] [PubMed]

5. Shetreat-Klein, M.; Shinnar, S.; Rapin, I. Abnormalities of joint mobility with autism spectrum disorders. *Brain Dev.* **2014**, *36*, 91–96. [CrossRef] [PubMed]

6. Eccles, J.A.; Iodice, V.; Dowell, N.G.; Owens, A.; Hughes, L.; Skipper, S.; Lycette, Y.; Humphries, K.; Harrison, N.A.; Mathias, C.J.; et al. Joint hypermobility and autonomic hyperactivity: Relevance to neurodevelopmental disorders. *J. Neurol. Neurosurg. Psychiatry* **2014**, *85*, e3. [CrossRef]

7. Demas, G.E.; Adamo, S.A.; French, S.S. Neuroendocrine-immune crosstalk in vertebrates and invertebrates: Implications for host defence. *Funct. Ecol.* **2011**, *25*, 29–39. [CrossRef]

8. Bukovsky, A. Immune system involvement in the regulation of ovarian function an augmentation of cancer. *Microsc. Res. Tech.* **2006**, *69*, 482–500. [CrossRef] [PubMed]

9. Matelski, L.; Van de Water, J. Risk factors in autism: Thinking outside the brain. *J. Autoimmun.* **2016**, *67*, 1–7. [CrossRef] [PubMed]

10. Careaga, M.; Rogers, S.; Handsen, R.L.; Amaral, D.G.; Van de Water, J.; Ashwood, P. Immune endophenotypes in children with autism spectrum disorder. *Biol. Psychiatry* **2017**, *81*, 434–441. [CrossRef] [PubMed]

11. Ingudomnukul, E.; Baron-Cohen, S.; Wheelwright, S.; Knickmeyer, R. Elevated rates of testosterone-related disorders in women with autism spectrum conditions. *Horm. Behav.* **2007**, *51*, 597–604. [CrossRef] [PubMed]

12. Xu, G.; Jing, J.; Bowers, K.; Liu, B.; Bao, W. Maternal diabetes and the risk of autism spectrum disorders in the offspring: A systematic review and meta-analysis. *J. Autism Dev. Disord.* **2014**, *44*, 766–775. [CrossRef] [PubMed]

13. Lee, B.K.; Arver, S.; Widman, L.; Gardner, R.M.; Magnusson, C.; Dalman, C.; Kosidou, K. Maternal hirsutism and autism spectrum disorders in offspring. *Autism Res.* **2017**, *10*, 1544–1546. [CrossRef] [PubMed]

14. Palomba, S.; Marotta, R.; Di Cello, A.; Russo, T.; Falbo, A.; Orio, F.; Tolino, A.; Zullo, F.; Esposito, R.; Sala, G.B.L. Pervasive developmental disorders in children of hyperandrogenic women with polycystic ovary syndrome: A longitudinal case-control study. *Clin. Endocrinol.* **2012**, *77*, 898–904. [CrossRef] [PubMed]

15. Cheung, I.; Vadas, P. A new disease cluster: Mast cell activation syndrome, postural orthostatic tachycardia syndrome, and Ehlers-Danlos syndrome. *J. Allergy Clin. Immunol.* **2015**, *135*, AB65. [CrossRef]

16. Castori, M.; Camerota, F.; Celletti, C.; Grammatico, P.; Padua, L. Ehlers-Danlos syndrome hypermobility type and the excess of affected females: Possible mechanisms and perspectives. *Am. J. Med. Genet. A* **2010**, *152*, 2406–2408. [CrossRef] [PubMed]

17. McIntosh, L.J.; Mallett, V.T.; Frahm, J.D.; Richardson, D.A.; Evans, M.I. Gynecologic disorders in women with Ehlers-Danlos syndrome. *J. Soc. Gynecol. Investig.* **1995**, *2*, 559–564. [CrossRef] [PubMed]

18. Hakim, A.J.; Grahame, R. A simple questionnaire to detect hypermobility: An adjunct to the assessment of patients with diffuse musculoskeletal pain. *Int. J. Clin. Pract.* **2003**, *57*, 163–166. [PubMed]

19. Lyall, K.; Van de Water, J.; Ashwood, P.; Hertz-Picciotto, I. Asthma and allergies in children with autism spectrum disorders: Results from the CHARGE study. *Autism Res.* **2015**, *8*, 567–574. [CrossRef] [PubMed]

20. Theoharides, T.C. Autism spectrum disorders and mastocytosis. *Int. J. Immunopath Pharmacol.* **2009**, *22*, 859–865. [CrossRef] [PubMed]

21. Comi, A.M.; Zimmerman, A.W.; Frye, V.H.; Law, P.A.; Peeden, J.N. Familial clustering of autoimmune disorders and evaluation of medical risk factors in autism. *J. Child Neurol.* **1999**, *14*, 388–394. [CrossRef] [PubMed]

22. Alberti, K.G.; Zimmet, P.; Shaw, J.; IDF Epidemiology Task Force Consensus Group. The metabolic syndrome—A new worldwide definition. *Lancet* **2005**, *366*, 1059–1062. [CrossRef]

23. Asthma and Allergy Foundation of America (AAFA). Allergy Facts and Figures. 2017. Available online: http://www.aafa.org/page/allergy-facts.aspx (accessed on 23 May 2017).

24. Akinbami, L.J.; Moorman, J.E.; Bailey, C.; Zahran, H.S.; King, M.; Johnson, C.A.; King, M.E.; Liu, X.; Moorman, J.E.; Zahran, H.S. Trends in asthma prevalence, health care use, and mortality in the United States, 2001–2010. *NCHS Data Brief* **2012**, *94*, 1–8.

25. Cooper, G.S.; Bynum, M.L.; Somers, E.C. Recent insights in the epidemiology of autoimmune diseases: Improved prevalence estimates and understanding of clustering of diseases. *J. Autoimmun.* **2009**, *33*, 197–207. [CrossRef] [PubMed]

26. Teele, D.W.; Klein, J.O.; Rosner, B. Epidemiology of otitis media during the first seven years of life in children in greater Boston: A prospective, cohort study. *J. Infect. Dis.* **1989**, *160*, 83–94. [CrossRef] [PubMed]

27. Monasta, L.; Ronfani, L.; Marchetti, F.; Montico, M.; Vecchi Brumatti, L.; Bavcar, A.; Grasso, D.; Barbiero, C.; Tamburlini, G. Burden of disease caused by otitis media: Systematic review and global estimates. *PLoS ONE* **2012**, *7*, e36226. [CrossRef] [PubMed]

28. Shi, J.B.; Fu, Q.L.; Zhang, H.; Cheng, L.; Wang, Y.J.; Zhu, D.D.; Lv, W.; Liu, S.X.; Li, P.Z.; Ou, C.Q.; et al. Epidemiology of chronic rhinosinusitis: Results from a cross-sectional survey in seven Chinese cities. *Allergy* **2015**, *70*, 533–539. [CrossRef] [PubMed]

29. Thong, B.Y.; Tan, T.C. Epidemiology and risk factors for drug allergy. *Br. J. Clin. Pharmacol.* **2011**, *71*, 684–700. [CrossRef] [PubMed]

30. Caress, S.M.; Steinemann, A.C. Prevalence of multiple chemical sensitivities: A population-based study in the southeastern United States. *Am. J. Pub. Health* **2004**, *94*, 746–747. [CrossRef]

31. Liebregts, T.; Adam, B.; Bredack, C.; Röth, A.; Heinzel, S.; Lester, S.; Downie–Doyle, S.; Smith, E.; Drew, P.; Talley, N.J.; et al. Immune activation in patients with irritable bowel syndrome. *Gastroenterology* **2007**, *132*, 913–920. [CrossRef] [PubMed]

32. Murialdo, G.; Magri, F.; Tamagno, G.; Ameri, P.; Camera, A.; Colnaghi, S.; Perucca, P.; Ravera, G.; Galimberti, C.A. Seizure frequency and sex steroids in women with partial epilepsy on antiepileptic therapy. *Epilepsia* **2009**, *50*, 1920–1926. [CrossRef] [PubMed]

33. De Paepe, A.; Malfait, F. The Ehlers-Danlos syndrome, a disorder with many faces. *Clin. Genet.* **2012**, *82*, 1–11. [CrossRef] [PubMed]

34. Matalliotakis, I.; Cakmak, H.; Matalliotakis, M.; Kappou, D.; Arici, A. High rate of allergies among women with endometriosis. *J. Obstet. Gyncaecol.* **2012**, *32*, 291–293. [CrossRef] [PubMed]

35. Sinali, N.; Cleary, S.D.; Ballweg, M.L.; Nieman, L.K.; Stratton, P. High rates of autoimmune and endocrine disorders, fibromyalgia, chronic fatigue syndrome and atopic diseases among women with endometriosis: A survey analysis. *Hum. Rep.* **2002**, *17*, 2715–2724. [CrossRef]

36. Collier, C.N.; Harper, J.C.; Cafardi, J.A.; Cantrell, W.C.; Wang, W.; Foster, K.W.; Elewski, B.E. The prevalence of acne in adults 20 years and older. *J. Am. Acad. Dermatol.* **2008**, *58*, 56–59. [CrossRef] [PubMed]

37. Münster, K.; Helm, P.; Schmidt, L. Secondary amenorrhea: Prevalence and medical contact—A cross-sectional study from a Danish county. *Br. J. Obstet. Gynaecol.* **1992**, *99*, 430–433. [CrossRef] [PubMed]

38. Mokdad, A.H.; Ford, E.S.; Bowman, B.A.; Dietz, W.H.; Vinicor, F.; Bales, V.S.; Marks, J.S. Prevalence of obesity, diabetes, and obesity-related health risk factors, 2001. *JAMA* **2003**, *289*, 76–79. [CrossRef] [PubMed]

39. Ju, H.; Jones, M.; Mishra, G. The prevalence and risk factors of dysmenorrhea. *Epidemiol. Rev.* **2014**, *36*, 104–113. [CrossRef] [PubMed]

40. Cramer, D.W.; Missmer, S.A. The epidemiology of endometriosis. *Ann. Acad. Sci.* **2002**, *955*, 11–22. [CrossRef]

41. Upadhyay, U.D.; Waddell, E.N.; Young, S.; Kerker, B.D.; Berger, M.; Matte, T.; Angell, S.Y. Prevalence, awareness, treatment, and control of high LDL cholesterol in New York City, 2004. *Prev. Chronic Dis.* **2010**, *7*, A61. [PubMed]

42. Escobar-Morreale, H.F.; Carmina, E.; Dewailly, D.; Gambineri, A.; Kelestimur, F.; Moghetti, P.; Pugeat, M.; Qiao, J.; Wijeyaratne, C.N.; Witchel, S.F.; et al. Epidemiology, diagnosis and management of hirsutism: A consensus statement by the Androgen Excess and Polycystic Ovary Syndrome Society. *Hum. Rep. Update* **2012**, *18*, 146–170. [CrossRef] [PubMed]

43. Centers for Disease Control and Prevention (CDC). Hypertension among Adults in the United States: National Health and Nutrition Examination Survey, 2011–2012. 2015. Available online: https://www.cdc.gov/nchs/products/databriefs/db133.htm (accessed on 5 July 2017).

44. Centers for Disease Control and Prevention (CDC). Infertility FAQs. 2017. Available online: https://www.cdc.gov/reproductivehealth/infertility/ (accessed on 5 July 2017).

45. Kotagasti, T. Prevalence of different menstrual irregularities in women with abnormal uterine bleeding (AUB)—An observational study. *Int. J. Curr. Res. Rev.* **2015**, *7*, 66.

46. Centers for Disease Control and Prevention (CDC). Obesity and Overweight. 2016. Available online: https://www.cdc.gov/nchs/fastats/obesity-overweight.htm (accessed on 5 July 2017).

47. Musmar, S.; Afaneh, A.; Mo'alla, H. Epidemiology of polycystic ovary syndrome: A cross sectional study of university students at An-Najah national university—Palestine. *Rep. Biol. Endocrinol.* **2013**, *11*, 47. [CrossRef] [PubMed]

48. Halbreich, U.; Borenstein, J.; Pearlstein, T.; Kahn, L.S. The prevalence, impairment, impact, and burden of premenstrual dysphoric disorder (PMS/PMDD). *Psychoneuroendocrinology* **2003**, *28*, 1–23. [CrossRef]

49. Silverberg, J.I.; Silverberg, N.B. Epidemiolgoy and extracutaneous comorbidities of severe acne in adolescence: A U.S. population-based study. *Br. J. Dermatol.* **2014**, *170*, 1136–1142. [CrossRef] [PubMed]

50. Zimmermann, A.; Bernuit, D.; Gerlinger, C.; Schaefers, M.; Geppert, K. Prevalence, symptoms and management of uterine fibroids: An international internet-based survey of 21,746 women. *BMC Womens Health* **2012**, *12*, 6. [CrossRef] [PubMed]

51. Warren-Ulanch, J.; Arslanian, S. Treatment of PCOS in adolescence. *Best Pract. Res. Clin. Endocrinol. Metab.* **2006**, *20*, 311–330. [CrossRef] [PubMed]

52. Rotterdam ESHRE, ASRM-Sponsored PCOS Consensus Workshop Group. Revised 2003 consensus on diagnostic criteria and long-term health risks related to polycystic ovary syndrome. *Fertil. Steril.* **2004**, *81*, 19–25.

53. Cutolo, M.; Sulli, A.; Capellino, S.; Villaggio, B.; Montagna, P.; Seriolo, B.; Montagna, P.; Seriolo, B.; Straub, R.H. Sex hormones influence on the immune system: Basic and clinical aspects in autoimmunity. *Lupus* **2004**, *13*, 635–638. [CrossRef] [PubMed]

54. Alexeef, S.E.; Yau, V.; Qian, Y.; Davignon, M.; Lynch, F.; Crawford, P.; Davis, R.; Croen, L.A. Medical conditions in the first years of life associated with future diagnosis of ASD children. *J. Autism Dev. Disord.* **2017**, *47*, 2067–2079. [CrossRef] [PubMed]

55. Mostafa, G.A.; Al-Ayadhi, L.Y. The possible relationship between allergic manifestations and elevated serum levels of brain specific auto-antibodies in autistic children. *J. Neuroimmunol.* **2013**, *261*, 77–81. [CrossRef] [PubMed]

56. Atladóttir, H.O.; Pedersen, M.G.; Thorsen, P.; Mortensen, P.B.; Deleuran, B.; Eaton, W.W.; Parner, E.T. Association of family history of autoimmune diseases and autism spectrum disorders. *Pediatrics* **2009**, *124*, 687–694. [CrossRef] [PubMed]

57. Lyall, K.; Ashwood, P.; Van de Water, J.; Hertz-Picciotto, I. Maternal immune-mediated conditions, autism spectrum disorders, and developmental delay. *J. Autism Dev. Disord.* **2014**, *44*, 1546–1555. [CrossRef] [PubMed]

58. Shulman, D.I.; Muhar, I.; Jorgensen, E.V.; Diamond, F.B.; Bercu, B.B.; Root, A.W. Autoimmune hyperthyroidism in prepubertal children and adolescents: Comparison of clinical and biochemical features at diagnosis and responses to medical therapy. *Thyroid* **1997**, *7*, 755–760. [CrossRef] [PubMed]

59. Felgentreff, K.; Siepe, M.; Kotthoff, S.; von Kodolitsch, Y.; Schachtrup, K.; Notarangelo, L.D.; Walter, J.E.; Ehl, S. Severe eczema and hyper-IgE in Loeys-Dietz syndrome—Contribution to new findings of immune dysfunction in connective tissue disorders. *Clin. Immunol.* **2014**, *150*, 43–50. [CrossRef] [PubMed]

60. Chung, B.H.; Lam, S.T.; Tong, T.M.; Li, S.Y.; Lun, K.S.; Chan, D.H.C.; Fok, S.F.S.; Or, J.S.F.; Smith, D.K.; Yang, W.; et al. Identification of novel FBN1 and TGFBR2 mutations in 65 probands with Mardan syndrome or Marfan-like phenotypes. *Am. J. Med. Genet. A* **2009**, *149*, 1452–1459. [CrossRef] [PubMed]

61. Neptune, E.R.; Frischmeyer, P.A.; Arking, D.E.; Myers, L.; Bunton, T.E.; Gayraud, B.; Ramirez, F.; Sakai, L.Y.; Dietz, H.C. Dysregulation of TGF-beta activation contributes to pathogenesis in Marfan syndrome. *Nat. Genet.* **2003**, *33*, 407–411. [CrossRef] [PubMed]

62. Matt, P.; Schoenhoff, F.; Habashi, J.; Holm, T.; Van Erp, C.; Loch, D.; Carlson, O.D.; Griswold, B.F.; Fu, Q.; De Backer, J.; et al. Circulating transforming growth factor-beta in Marfan syndrome. *Circulation* **2009**, *120*, 526–532. [CrossRef] [PubMed]

63. Holm, T.M.; Habashi, J.P.; Doyle, J.J.; Bedja, D.; Chen, Y.; van Erp, C.; Lindsay, M.E.; Kim, D.; Schoenhoff, F.; Cohn, R.D.; et al. Noncanonical TGF-beta signaling contributes to aortic aneurysm progression in Marfan syndrome mice. *Science* **2011**, *332*, 358–361. [CrossRef] [PubMed]

64. Kaartinen, V.; Warburton, D. Fibrillin controls TGF-beta activation. *Nat. Genet.* **2003**, *33*, 331–332. [CrossRef] [PubMed]

65. Frischmeyer-Guerrerio, P.A.; Guerrerio, A.l.; Oswald, G.; Chichester, K.; Myers, L.; Halusha, M.K.; Oliva-Hemker, M.; Wood, R.A.; Dietz, H.C. TGF-beta receptor mutations impose a strong predisposition for human allergic disease. *Sci. Transl. Med.* **2013**, *5*, 195ra94. [CrossRef] [PubMed]

66. Malfait, F.; Francomano, C.; Byers, P.; Belmont, J.; Berglund, B.; Black, J.; Bloom, L.; Bowen, J.M.; Brady, A.F.; Burrows, N.P.; et al. The 2017 international classification of the Ehlers-Danlos syndromes. *Am. J. Med. Genet. C Semin. Med. Genet.* **2017**, *175*, 8–26. [CrossRef] [PubMed]

67. Ashwood, P.; Enstrom, A.; Krakowiak, P.; Hertz-Picciotto, I.; Hansen, R.L.; Croen, L.A.; Ozonoff, S.; Pessah, I.N.; Van de Water, J. Decreased transforming growth factor beta1 in autism: A potential link between immune dysregulation and impairment in clinical behavioral outcomes. *J. Neuroimmunol.* **2008**, *204*, 149–153. [CrossRef] [PubMed]

68. Okada, K.; Hashimoto, K.; Iwata, Y.; Nakamura, K.; Tsujii, M.; Tsuchiya, K.J.; Sekine, Y.; Suda, S.; Suzuki, K.; Sugihara, G.I.; et al. Decreased serum levels of transforming growth factor-β1 in patients with autism. *Prog. Neuropsychopharmacol. Biol. Psychiatry* **2007**, *31*, 187–190. [CrossRef] [PubMed]

69. Castori, M.; Morlino, S.; Dordoni, C.; Celletti, C.; Camerota, F.; Ritelli, M.; Morrone, A.; Venturini, M.; Grammatico, P. Gynecologic and obstetric implications of the joint hypermobility syndrome (a.k.a. Ehlers-Danlos syndrome hypermobility type) in 82 Italian patients. *Am. J. Med. Genet. A* **2012**, *158*, 2176–2182. [CrossRef] [PubMed]

70. Muldowney, P.T.K. *Living Life to the Fullest with Ehlers-Danlos Syndrome: Guide to Living a Better Quality Life While Having EDS*; Outskirts Press: Denver, CO, USA, 2015.

71. Talwar, R.M.; Wong, B.S.; Svoboda, K.; Harper, R.P. Effects of estrogen on chondrocyte proliferation and collagen synthesis in skeletally mature articular cartilage. *J. Oral. Maxillofac. Surg.* **2006**, *64*, 600–609. [CrossRef] [PubMed]

72. Henneman, D.H. Effect of estrogen on in vivo and in vitro collagen biosynthesis and maturation in old and young female guinea pigs. *Endocrinology* **1968**, *83*, 678–690. [CrossRef] [PubMed]

73. Hansen, M.; Kongsgaard, M.; Holm, L.; Skovgaard, D.; Magnusson, S.P.; Qvortrup, K.; Larsen, J.O.; Aagaard, P.; Dahl, M.; Serup, A.; et al. Effect of estrogen on tendon collagen synthesis, tendon structural characteristics, and biomechanical properties in postmenopausal women. *J. Appl. Physiol.* **2009**, *106*, 1385–1393. [CrossRef] [PubMed]

74. Salem, M.L. Estrogen, a double-edged sword: Modulation of TH1- and TH2-mediated inflammations by differential regulation of TH1/TH2 cytokine production. *Curr. Drug. Targets Inflamm. Allergy* **2004**, *3*, 97–104. [CrossRef] [PubMed]

75. Zierau, O.; Zenclussen, A.C.; Jensen, F. Role of female sex hormones, estradiol and progesterone, in mast cell behavior. *Front. Immunol.* **2012**, *3*, 169. [CrossRef] [PubMed]

76. Urb, M.; Sheppard, D.C. The role of mast cells in the defence against pathogens. *PLoS Pathog.* **2012**, *8*, e1002619. [CrossRef] [PubMed]

77. Hawse, J.R.; Subramaniam, M.; Ingle, J.N.; Oursler, M.J.; Rajamannan, N.M.; Spelberg, T.C. Estrogen-TGF-beta cross-talk in bone and other cell types: Role of TIEG, Runx2, and other transcription factors. *J. Cell. Biochem.* **2012**, *103*, 383–392. [CrossRef] [PubMed]

78. Afrin, L.B. *Never Bet Against Occam*; Sisters Media, LLC: Bethesda, MD, USA, 2016.

79. Seneviratne, S.L.; Maitland, A.; Afrin, L. Mast cell disorders in Ehlers-Danlos syndrome. *Am. J. Med. Genet. C Semin. Med. Genet.* **2017**, *175*, 226–236. [CrossRef] [PubMed]

80. De Wandele, I.; Rombaut, L.; Leybaert, L.; Van de Borne, P.; De Backer, T.; Malfait, F.; De Paepe, A.; Calders, P. Dysautonomia and its underlying mechanisms in the hypermobility type of Ehlers-Danlos syndrome. *Semin. Arthritis Rheum.* **2014**, *44*, 93–100. [CrossRef] [PubMed]

81. Cairns, J.A.; Walls, A.F. Mast cell tryptase stimulates the synthesis of type I collagen in human lung fibroblasts. *J. Clin. Investig.* **1997**, *99*, 1313–1321. [CrossRef] [PubMed]

82. Yirmiya, R.; Goshen, I. Immune modulation of learning, memory, neural plasticity and neurogenesis. *Brain Behav. Immun.* **2011**, *25*, 181–213. [CrossRef] [PubMed]

83. Martínez-Cerdeño, V.; Noctor, S.C.; Kriegstein, A.R. Estradiol stimulates progenitor cell division in the ventricular and subventricular zones of the embryonic neocortex. *Eur. J. Neurosci.* **2011**, *24*, 3475–3488. [CrossRef] [PubMed]

84. Sato, K.; Akaishi, T.; Matsuki, N.; Ohno, Y.; Nakazawa, K. β-estradiol induces synaptogenesis in the hippocampus by enhancing brain-derived neurotrophic factor release from dentate gyrus granule cells. *Brain Res.* **2007**, *1150*, 108–120. [CrossRef] [PubMed]

85. Connolly, A.M.; Chez, M.; Streif, E.M.; Keeling, R.M.; Golumbek, P.T.; Kwon, J.M.; Riviello, J.J.; Robinson, R.G.; Neuman, R.J.; Deuel, R.M.K. Brain-derived neurotrophic factor and autoantibodies to neural antigens in sera of children with autistic spectrum disorders, Landau-Kleffner syndrome, and epilepsy. *Biol. Psychiatry* **2006**, *59*, 354–363. [CrossRef] [PubMed]

86. D'Amelio, M.; Ricci, I.; Sacco, R.; Liu, X.; D'Agruma, L.; Muscarella, L.A.; Guarnieri, V.; Militerni, R.; Bravaccio, C.; Elia, M.; et al. Paraoxonase gene variants are associated with autism in North America, but not in Italy: Possible regional specificity in gene-environment interactions. *Mol. Psychiatry* **2005**, *10*, 1006–1016. [CrossRef] [PubMed]

87. Cederlöf, M.; Larsson, H.; Lichtenstein, P.; Almqvist, C.; Serlachius, E.; Ludvigsson, J.F. Nationwide population-based cohort study of psychiatric disorders in individuals with Ehlers-Danlos syndrome or hypermobility syndrome and their siblings. *BMC Psychiatry* **2016**, *16*, 207. [CrossRef] [PubMed]

88. National Organization for Rare Disorders (NORD). Ehlers Danlos Syndrome. 2017. Available online: https://rarediseases.org/rare-diseases/ehlers-danlos-syndrome/#affected-populations (accessed on 18 June 2017).

89. Lyons, J.J.; Sun, G.; Stone, K.D.; Nelson, C.; Wisch, L.; O'Brien, M.; Jones, N.; Lindsley, A.; Komarow, H.D.; Bai, Y.; et al. Mendelian inheritance of elevated serum tryptase associated with atopy and connective tissue abnormalities. *J. Allergy Clin. Immunol.* **2014**, *133*, 1471–1474. [CrossRef] [PubMed]

90. Vezzani, A.; Moneta, D.; Richichi, C.; Aliprandi, M.; Burrows, S.J.; Ravizza, T.; Perego, C.; De Simoni, M.G. Functional role of inflammatory cytokines and antiinflammatory molecules in seizures and epileptogenesis. *Epilepsia* **2002**, *43*, 30–35. [CrossRef] [PubMed]

behavioral sciences

MDPI

Article

Alterations in the MicroRNA of the Blood of Autism Spectrum Disorder Patients: Effects on Epigenetic Regulation and Potential Biomarkers

Tamara da Silva Vaccaro [1], Julia Medeiros Sorrentino [1], Sócrates Salvador [2], Tiago Veit [3], Diogo Onofre Souza [1] and Roberto Farina de Almeida [1,*]

[1] Institute of Health's Basic Science, Department of Biochemistry, Federal University of Rio Grande do Sul, Porto Alegre 90035-000, RS, Brazil; tamara.genetica@gmail.com (T.d.S.V.); juliamsorrentino@gmail.com (J.M.S.); diogo.bioq@gmail.com (D.O.S.)

[2] Pediatric Neurology Center, Porto Alegre Clinical Hospital (HCPA), Federal University of Rio Grande do Sul, Porto Alegre 90035-903, RS, Brazil; socratessalvador@gmail.com

[3] Institute of Health's Basic Science, Department of Microbiology, Immunology and Parasitology, Federal University of Rio Grande do Sul, Porto Alegre 90035-190, RS, Brazil; tiagoveit@terra.com.br

* Correspondence: almeida_rf@yahoo.com.br

Received: 12 June 2018; Accepted: 11 August 2018; Published: 15 August 2018

Abstract: Aims: Autism spectrum disorder (ASD) refers to a group of heterogeneous brain-based neurodevelopmental disorders with different levels of symptom severity. Given the challenges, the clinical diagnosis of ASD is based on information gained from interviews with patients' parents. The heterogeneous pathogenesis of this disorder appears to be driven by genetic and environmental interactions, which also plays a vital role in predisposing individuals to ASD with different commitment levels. In recent years, it has been proposed that epigenetic modifications directly contribute to the pathogenesis of several neurodevelopmental disorders, such as ASD. The microRNAs (miRNAs) comprises a species of short noncoding RNA that regulate gene expression post-transcriptionally and have an essential functional role in the brain, particularly in neuronal plasticity and neuronal development, and could be involved in ASD pathophysiology. The aim of this study is to evaluate the expression of blood miRNA in correlation with clinical findings in patients with ASD, and to find possible biomarkers for the disorder. **Results:** From a total of 26 miRNA studied, seven were significantly altered in ASD patients, when compared to the control group: miR34c-5p, miR92a-2-5p, miR-145-5p and miR199a-5p were up-regulated and miR27a-3p, miR19-b-1-5p and miR193a-5p were down-regulated in ASD patients. **Discussion:** The main targets of these miRNAs are involved in immunological developmental, immune response and protein synthesis at transcriptional and translational levels. The up-regulation of both miR-199a-5p and miR92a-2a and down-regulation of miR-193a and miR-27a was observed in AD patients, and may in turn affect the SIRT1, HDAC2, and PI3K/Akt-TSC:mTOR signaling pathways. Furthermore, MeCP2 is a target of miR-199a-5p, and is involved in Rett Syndrome (RTT), which possibly explains the autistic phenotype in male patients with this syndrome.

Keywords: ASD; microRNA; SIRT1; HDAC2; PI3K/Akt-TSC:mTOR; MeCP2

---

1. Introduction

Autism spectrum disorder (ASD) refers to a group of heterogeneous brain-based neurodevelopmental disorders with different levels of symptom severity in two core domains: impairment in communication and social interaction; and repetitive and stereotypic patterns of behavior [1]. Given the challenges, the clinical diagnosis of ASD is based on information gained from interviews with patients' parents, in accordance with the Diagnostic and Statistical Manual of Mental Disorders, Fifth Edition, Text Revision (DSMV-TR). In April 2018 an important study from the Autism and Developmental Disabilities Monitoring Network was published by the U.S. Department of Health and Human Services/Centers for Disease Control that estimates the prevalence of ASD among U.S. children aged 8 years old. The results obtained revealed that one in 59 children aged 8 years, from multiple communities, presented ASD [1]. Multiple lines of evidence suggest that ASD is genetic in origin, with most data supporting a polygenic model [2,3]. However, except for the GWAS study [4], which demonstrate that although autism possesses a complex genetic architecture, common variations are found, and other genetic studies have been quite successful in identifying suitable candidate genes for ASD. The heterogeneous pathogenesis of this disorder appears to be driven by genetic and environmental interactions, which also play a vital role in predisposing individuals to ASD with different commitment levels [5]. In recent years, it has been proposed that epigenetic modifications directly contribute to the pathogenesis of several neurodevelopmental disorders, such as ASD [6–8]. Epigenetics modifiers act at the interface of genes targeting different mechanisms: histone modifications, DNA methylation, chromatin remodeling or noncoding RNA (microRNAs); controlling heritable changes in gene expression without changing the DNA sequence [9,10].

It has been proposed that epigenetic machinery was closely related with neuronal development disorders by several pathways influencing cell cycle regulation, development and axon guidance, dendritic spine development and function, actin cytoskeleton regulation and protein synthesis regulation [11]. The microRNAs (miRNAs) comprise a species of short noncoding RNA (approximately 21 nucleotides) that regulate gene expression post-transcriptionally, by interacting with specific mRNAs, usually at the $3'$ untranslated region (UTR), through partial sequence complementation, resulting in mRNA degradation or repression of translation [12]. In this way, a variety of cellular processes could be affected, as cellular differentiation, metabolism and apoptosis [13]. miRNAs have an essential functional role in the brain, particularly in neuronal plasticity and neuronal development [14]. Each miRNA binds multiple targets in several mature mRNA species, mediating several biological functions, including physiological neuronal gene expression and also several pathological processes, as were previously described in some brain disorders [15,16]. The expression of many miRNAs is dynamically regulated during brain development, neurogenesis, the neuronal maturation mechanism [14], and an important signaling pathway involved in all of these processes is the tuberous sclerosis complex-mammalian target of rapamycin (TSC-mTOR). For example, a significant impairment to this pathway has been observed in some disorders related to autism, including Tuberous Sclerosis, a disease characterized by mutations in tuberous sclerosis proteins (TSC) TSC1 or TSC2 genes [17].

Recently, studies have revealed associations between the immune system and the central nervous system in ASD development, hypothesizing that early neuroimmune disturbances during embryogenesis could persist throughout an individual's lifetime. Considering that miRNAs can pass into the bloodstream from cells or tissues and organs [18], it seems reasonable to suppose that changes in miRNA levels in blood could reflect direct changes occurred in the Central Nervous System and lymphoid organs. Thus, in the present study, we evaluated the expression of miRNA in the blood of ASD patients and analyzed the repression targets of these miRNAs, correlating them with biochemical pathways that may be deregulated in ASD. We propose that the miRNA expression profile may be used as a clinical marker for the diagnosis or prognosis of the disorder, and that epigenetic changes may help in understanding the disease.

2. Materials and Methods

2.1. Subjects

This study was approved by the Ethics Committee of Hospital de Clínicas de Porto Alegre (HCPA) and the subjects' parents provided informed consent before inclusion in the study. Eleven patients attending an outpatient clinic of the HCPA were included in the study, following a semi-structured interview. Clinical diagnosis was confirmed by criteria defined by the Diagnostic and Statistical Manual of Mental Disorders V and Autism Screening Questionnaire. Seven ASD male patients with a mean age of 7.5 years (sd 2.5) and four non-ASD male controls with a mean age of 7.5 years (sd 2.5) was carried out. The ASD subjects were enrolled in the same autism severity group by two well-validated clinical tests: the Childhood Autism Rating Scale (CARS) [19] and Autism Diagnostic Observation Schedule-Generic (ADOS-G) [20]. Exclusion criteria were as follows: (a) comorbidities such as chromosomal syndromes; (b) genetic or metabolic disease. Non-ASD male children were included after clinical diagnosis that excluded: (a) presence of psychiatric disorder; (b) presence of ASD patients in family; (c) presence of chromosomal syndrome; (d) genetic or metabolic disease.

2.2. Quantitative Real Time Polymerase Chain Reaction (RT-qPCR)

MicroRNA profile evaluations were evaluated from peripheral blood samples were obtained from patient and immediately mixed with a three-fold volume of Trizol (Invitrogen) for total RNA extraction.

Considering some common aspects among the different disorders that affect the Central Nervous System, the 26 candidate miRNAs evaluated in this study, were based in previous studies in the literature conducted with patients with sleep disturbances [21] or who had some neurodegenerative diseases [22,23].

miRNA quantification was carried out by quantitative RT-qPCR using a stem-loop RT-PCR technique, as previously described [24]. Complementary DNA (cDNA) was synthesized from mature miRNA using a reverse transcriptase reaction containing 2 μg of total RNA, 1 μL of 10 mM dNTP mix (Invitrogen, Waltham, MA, USA), 3 μL of stem loop RT primer mix, 4 μL M-MLV reverse transcriptase 5X reaction buffer (Invitrogen), 2 μL of 0.1 M DTT (Invitrogen), 1 μL of RNase inhibitor (Invitrogen), 1.0 μL of M-MLV reverse transcriptase (Invitrogen), and sterile distilled water to a final volume of 20 μL. The synthesis of cDNA was completed after a sequence of four incubations at 16 °C for 30 min, 42 °C for 30 min, 85 °C for 5 min and 4 °C for 10 min.

The quantitative PCR mix was made up with 12 μL of cDNA (1:33), 1.0 μL of specific miRNA forward and universal reverse (10 μM) primers (as detailed in Table 1), 0.5 μL of 10 μM dNTP mix, 2.5 μL of 10X PCR buffer (Invitrogen), 1.5 μL of 50 mM MgCl2 (Invitrogen), 2.4 μL of 1X Sybr Green (Molecular Probes, Eugene, OR, USA), 0.1 μL of Platinum Taq DNA Polymerase (Invitrogen), and sterile distilled water to a final volume of 24 μL. The fluorescence of Sybr Green was used to detect amplification, estimate Ct values, and to determine specificity after melting curve analysis. PCR cycling conditions were standardized to 95 °C for 5 min, followed by 40 cycles at 95 °C for 10 s, 60 °C for 10 s, and 72 °C for 10 s. After the main amplification, sample fluorescence was measured from 60 °C to 95 °C, with an increasing ramp of 0.3 °C each, in order to obtain the denaturing curve of the amplified products to assure their homogeneity after peak detection, and Tm (melting temperature) estimation using data obtained from an Applied Biosystems StepOne System (Lincoln Centre Drive Foster City, CA, USA). After the main amplification step, sample fluorescence was measured at temperatures from 55–99 °C, with an increasing ramp of 0.1 °C, in order to obtain the denaturing curve of the amplified products and to assure their homogeneity after peak detection and Tm estimation using data obtained from Applied Biosystems StepOne Plus.

The relative expression was obtained in triplicate using the $2^{-\Delta\Delta Ct}$ method, where the Crossing threshold (Ct) values of the target samples are subtracted from the average Ct values of the standard or control samples. The use of $2^{-\Delta\Delta Ct}$ is adequate, as the amount of RNA among the different blood samples to produce the cDNAs did not differ significantly and produced similar Ct values among the samples for 4 different miRNA used in the initial screening and evaluated by Genorm software. The Genorm analysis was used to assess the variance in the expression levels among the miRNA and pairwise comparisons. This resulted in 4 control miRNA, the most stable ones, for every pairwise comparison, serving as normalizers to evaluate the relative expression of miRNA. The RT-qPCR results were analyzed by Genorm algorithm to assess the variance in expression levels of the miRNA studied.

This program performed a scan of the present miRNA from groups compared two by two each time. Then, the expression stabilities of the set of miRNA were evaluated. All miRNA were ranked according to their stability value. A pairwise variation analysis was performed by Genorm to determine the number of miRNA required for accurate normalization and to identify which miRNA could be used as internal control.

Table 1. Information about miRNA evaluated. ID, Chromosome localization, Accession miRBase (www.mirbase.org), Mature Sequence and Forward Primer.

ID	Chromosome	Accession miRBase	Mature Sequence	Forward Primer
hsa-miR-19b-1-5p	13	MI0000074	AGUUUUGCAGGUUUGCAUCCAGC	AGTTTTGCAGGTTTGCATCCAGC
hsa-miR-24-2-5p	19	MI0000081	UGCCUACUGAGCUGAAACACAG	TGCCTACTGAGCTGAAACACAG
hsa-miR-25-3p	7	MI0000082	CAUUGCACUUGUCUCGGUCUGA	CATTGCACTTGTCTCGGTCTGA
hsa-miR-27a-3p	19	MI0000085	UUCACAGUGGCUAAGUUCCGC	TTCACAGTGGCTAAGTTCCGC
hsa-miR-29b-2-5p	1	MI0000107	CUGGUUUCACAUGGUGGCUUAG	CTGGTTTCACATGGTGGCTTAG
hsa-miR-31-5p	9	MI0000089	AGGCAAGAUGCUGGCAUAGCU	AGGCAAGATGCTGGCATAGCT
hsa-miR-34a-5p	1	MI0000268	UGGCAGUGUCUUAGCUGGUUGU	TGGCAGTGTCTTAGCTGGTTGT
hsa-miR-34c-5p	11	MI0000743	AGGCAGUGUAGUUAGCUGAUUGC	AGGCAGTGTAGTTAGCTGATTGC
hsa-miR-92a-2-5p	X	MI0000094	GGGUGGGGAUUUGUUGCAUUAC	GGGTGGGGATTTGTTGCATTAC
hsa-miR-99a-5p	21	MI0000101	AACCCGUAGAUCCGAUCUUGUG	AACCCGTAGATCCGATCTTGTG
hsa-miR-125a-5p	19	MI0000469	UCCCUGAGACCCUUUAACCUGUGA	TCCCTGAGACCCTTTAACCTGTGA
hsa-miR-125b-1-3p	11	MI0000446	ACGGGUUAGGCUCUUGGGAGCU	ACGGGTTAGGCTCTTGGGAGCT
hsa-miR-125b-2-3p	21	MI0000470	UCACAAGUCAGGCUCUUGGGAC	TCACAAGTCAGGCTCTTGGGAC
hsa-miR-145-5p	5	MI0000461	GUCCAGUUUUCCCAGGAAUCCCU	GTCCAGTTTTCCCAGGAATCCCT
hsa-miR-181b-5p	1	MI0000270	AACAUUCAUUGCUGUCGGUGGGU	AACATTCATTGCTGTCGGTGGGT
hsa-miR-191-5p	3	MI0000465	CAACGGAAUCCCAAAAGCAGCUG	CAACGGAATCCCAAAAGCAGCTG
hsa-miR-193a-5p	17	MI0000487	UGGGUCUUUGCGGGCGAGAUGA	TGGGTCTTTGCGGGCGAGATGA
hsa-miR-193b-3p	16	MI0003137	AACUGGCCCUCAAAGUCCCGCU	AACTGGCCCTCAAAGTCCCGCT
hsa-miR-198	3	MI0000240	GGUCCAGAGGGGAGAUAGGUUC	GGTCCAGAGGGGAGATAGGTTC
hsa-miR-199a-5p	19	MI0000242	CCCAGUGUUCAGACUACCUGUUC	CCCAGTGTTCAGACTACCTGTTC
hsa-miR-210-3p	11	MI0000286	CUGUGCGUGUGACAGCGGCUGA	CTGTGCGTGTGACAGCGGCTGA
hsa-miR-214-3p	1	MI0000290	ACAGCAGGCACAGACAGGCAGU	ACAGCAGGCACAGACAGGCAGT
hsa-miR-221-3p	X	MI0000298	AGCUACAUUGUCUGCUGGGUUUC	AGCTACATTGTCTGCTGGGTTTC
hsa-miR-222-3p	X	MI0000299	AGCUACAUCUGGCUACUGGGU	AGCTACATCTGGCTACTGGGT
hsa-miR-339-5p	7	MI0000815	UCCCUGUCCUCCAGGAGCUCACG	TCCCTGTCCTCCAGGAGCTCACG
hsa-miR-370-3p	14	MI0000778	GCCUGCUGGGGUGGAACCUGGU	GCCTGCTGGGGTGGAACCTGGT

2.3. Statistical Analysis

To assess whether ASD patients and controls were homogeneous for age and gender the Chi-squared test was used. Statistical analysis of the relative expression values obtained for each miRNA in the experimental group was performed by Student's *t*-test implemented using the SPSS Statistics 17 software. In order to compare the expression levels between the two experimental groups, the Waller–Duncan and Tukey HSD tests were performed with SPSS 17, with identical group discrimination and similar probability values.

Each altered miRNA has a cluster of validated targets, which were predicted for at least two available programs for experimental assays (miRBase).

The protein clusters were formed by the STRING 10 software using an automatic force-directed layout algorithm that orders the nodes in the network. The algorithm works iteratively trying to position the nodes apart from each other with a "preferred distance" proportional to the String global score (https://string-db.org/). All nucleotide sequences were screened and evaluated through miRBase and, after the ready profile, miRBase data banks were searched for altered microRNA targets (www.mirbase.org).

3. Results

From a total of 26 miRNAs evaluated (Table 1), seven were statistically altered in ASD patients in comparison with control group. Specifically, miR34c-5p (Figure 1; $p = 0.0068$), miR-145-5p (Figure 2; $p = 0.0099$), miR92a-2-5p (Figure 3; $p = 0.0026$) and miR199a-5p (Figure 4; $p = 0.047$) were up-regulated, while miR19b-1-5p (Figure 5; $p = 0.0184$), miR27a-3p (Figure 6; $p = 0.0001$) and miR193a-5p (Figure 7; $p = 0.001$) were down-regulated comparing the ASD patients with the control group. Additionally, the validated targets of the seven altered miRNA are shown in protein clusters (Figures 1b–7b), except for the miR92a-2-5p, which does not have validated targets in *Homo sapiens*. The results for Genorm revealed ubiquitous and stably expressed normalization by RT-qPCR: miR125a-5p, miR181b-5p, miR125b-2-3p, miR198. Thus, each alerted miRNA was calculated from 4 normalizing miRNA.

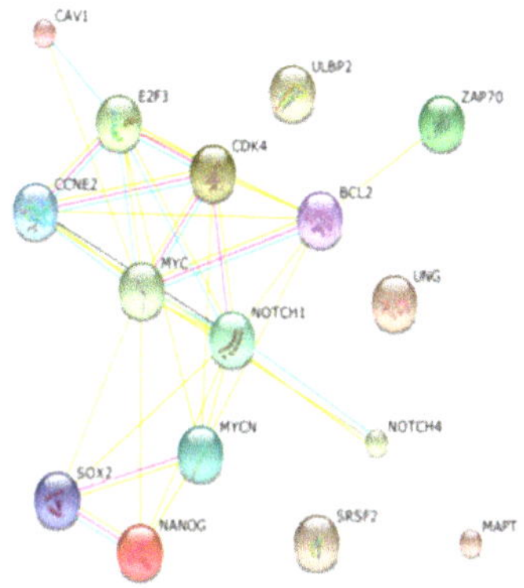

Figure 1. (a) Scatter plot of differential relative expression of miR34c in peripheral blood of ASD subjects compared to control subjects. Results expressed as mean ± standard error. *t*-Test analysis; * $p < 0.05$. (b) mRNA validates targets for miR34c selected in miRBase and respective proteins clusters that can be involved in ASD. Proteins involved in epigenetic regulation: NANOG (Nanog homeobox); NOTCH1 (notch 1); SOX2 (SRY (sex determining region Y)-box 2); SRSF2 (serine/arginine-rich splicing factor 2); NOTCH4 (notch 4); E2F3 (E2F transcription factor); MYCN (v-myc myelocytomatosis viral related oncogene, neuroblastoma derived); MYC (v-myc myelocytomatosis viral oncogene homolog). Proteins involved in cell cycle: CCNE2 (cyclin E2); BCL2 (B-cell CLL/lymphoma 2). Proteins involved in immunological regulation: ZAP70 (zeta-chain (TCR) associated protein kinase 70kDa); ULBP2 (UL16 binding protein 2); CDK4 (cyclin-dependent kinase 4); CAV1 (caveolin 1). Protein associated with cytoskeleton stabilization in neuronal cell: MAPT (microtubule-associated protein tau (776 aa)). Protein involved in DNA repair: UNG (uracil-DNA glycosylase).

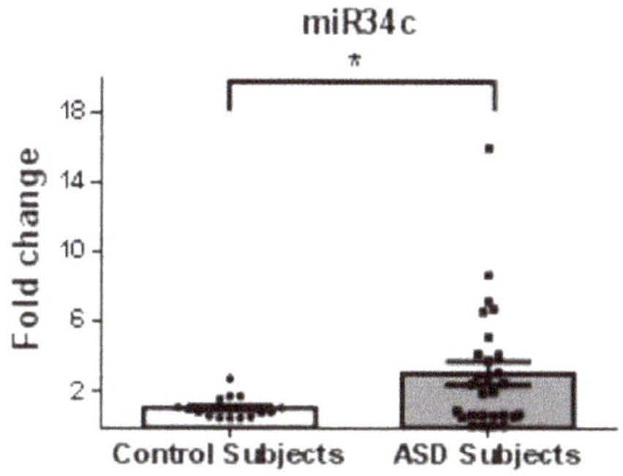

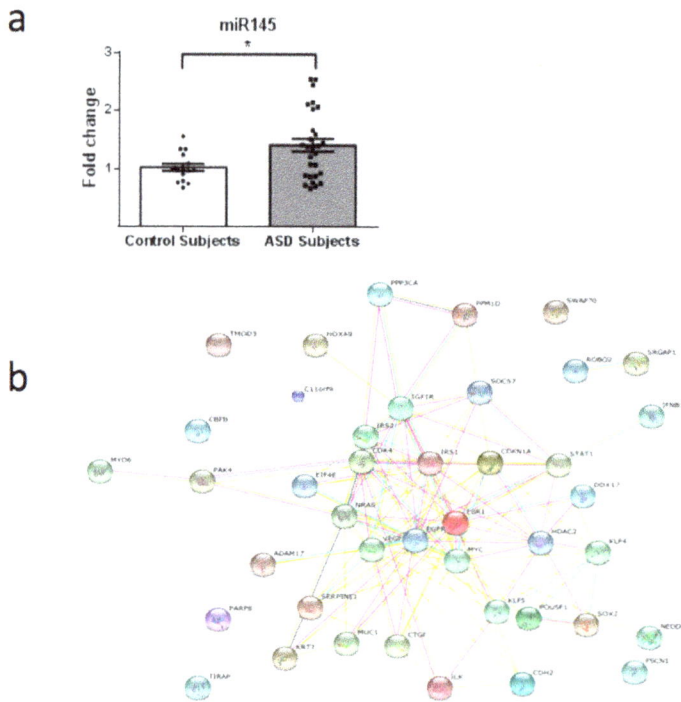

Figure 2. (**a**) Scatter plot of differential relative expression of miR145 in peripheral blood of ASD subjects compared to control subjects. Results expressed as mean ± standard error. *t*-Test analysis; * $p < 0.05$. (**b**) mRNA validates targets for miR145 selected in miRBase and respective proteins clusters that can be involved in ASD. Proteins involved in epigenetic regulation: ERS1 (estrogen receptor 1); POU5F1 (POU class 5 homeobox 1 (360 aa)); C11orf9 (chromosome 11 open reading frame 9); PARP8 (poly (ADP-ribose) polymerase family, member 8 (854 aa)); SOX2 (SRY (sex determining region Y)-box 2); HOX9 (homeobox A9); STAT1 (signal transducer and activator of transcription 1); KLF4 (Kruppel-like factor 4); KLF5 (Kruppel-like factor 5); NEDD9 (neural precursor cell expressed); DDX17 (DEAD (Asp-Glu-Ala-Asp) box helicase 17); EIF4E (eukaryotic translation initiation factor 4E); CBFB (core-binding factor, beta subunit); HDAC2 (histone deacetylase 2). Proteins involved in cell cycle: CDKN1A (cyclin-dependent kinase inhibitor 1A (p21, Cip1)); CDK4 (cyclin-dependent kinase 4); MYC (v-myc myelocytomatosis viral oncogene homolog); PPM1D (protein phosphatase, Mg^{2+}/Mn^{2+} dependent, 1D); KRT7 (keratin 7). Proteins involved in immunological regulation: IFNB1 (interferon beta 1 fibroblast); TIRAP (toll-interleukin 1 receptor (TIR) domain containing adaptor protein); SOCS7 (suppressor of cytokine signaling 7); ADAM17 (ADAM metallopeptidase domain 17). Proteins involved in insulin metabolism: IGF1R (insulin-like growth factor 1 receptor), IRS1 (insulin receptor substrate 1); IRS2 (insulin receptor substrate 2). Proteins associated with cytoskeleton and cell migration: SWAP70 (SWAP switching B-cell complex 70kDa subunit); ILK (integrin-linked kinase); MYO6 (myosin VI); FSCN1 (fascin homolog 1, actin-bundling protein); ROBO2 (roundabout, axon guidance receptor, homolog 2); CDH2 (cadherin 2, type 1, N-cadherin (neuronal)), TMOD3 (tropomodulin 3 (ubiquitous)); SRGAP1 (SLIT-ROBO Rho GTPase activating protein 1); PAK4 (p21 protein (Cdc42/Rac)-activated kinase 4). Others: SERPINE1 (serpin peptidase inhibitor, clade E (nexin, plasminogen activator inhibitor type 1)); EGFR (epidermal growth factor receptor); NRAS (neuroblastoma RAS viral (v-ras) oncogene homolog); VEGFA (vascular endothelial growth factor A); PPP3CA (protein phosphatase 3, catalytic subunit); CTGF (connective tissue growth factor); MUC (mucin 1).

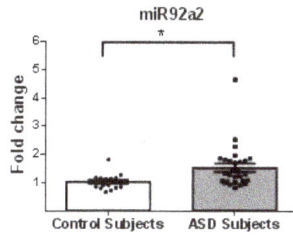

Figure 3. Scatter plot of differential relative expression of miR92a2 in peripheral blood of ASD subjects compared to control subjects. Results expressed as mean ± standard error. *t*-Test analysis; * $p < 0.05$. miR92a2 does not have validated targets in Homo sapiens.

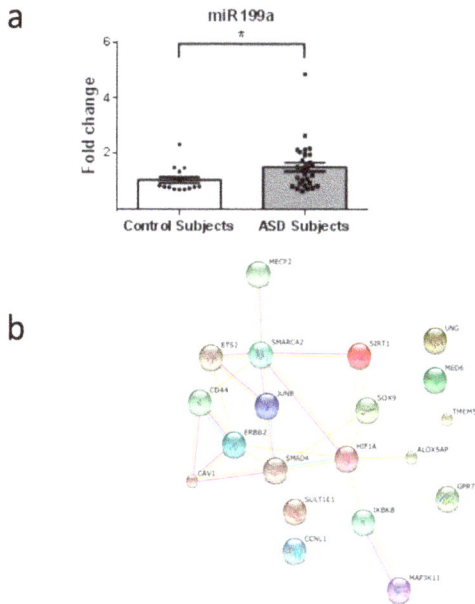

Figure 4. (**a**) Scatter plot of differential relative expression of miR199a in peripheral blood of ASD subjects compared to control subjects. Results expressed as mean ± standard error. *t*-Test analysis; * $p < 0.05$. (**b**) mRNA validates targets for miR199a selected in miRBase and respective proteins clusters that can be involved in ASD. Proteins involved in epigenetic regulation: SIRT1 (sirtuin 1); SOX9 (SRY (sex determining region Y)-box 9); MED6 (mediator complex subunit 6); SMARCA2 (SWI/SNF related, matrix associated, actin dependent regulator of chromatin, subfamily a, member 2); CCNL1 (cyclin L1); JUNB (jun B proto-oncogene); HIF1A (hypoxia inducible factor 1, alpha subunit); ETS2 (v-ets erythroblastosis virus E26 oncogene homolog 2); MECP2 (methyl-CpG binding protein 2). Proteins involved in immunological regulation: CAV1 (caveolin 1); SMAD4 (SMAD family member 4); ALOX5AP (arachidonate 5-lipoxygenase-activating protein); CD44 (CD44 molecule); IKBKB (inhibitor of kappa light polypeptide gene enhancer in B-cells). Protein associated with cytoskeleton and cell migration: ERBB2 (v-erb-b2 erythroblastic leukemia viral oncogene homolog 2). Protein involved in DNA repair: UNG (uracil-DNA glycosylase). Protein involved in strogen metabolim: SULT1E1 (sulfotransferase family 1E, estrogen-preferring, member 1). Others: MAP3K11 (mitogen-activated protein kinase kinase kinase 11); TMEM54 (transmembrane protein 54); GPR78 (G protein-coupled receptor 78; Orphan receptor).

a

b

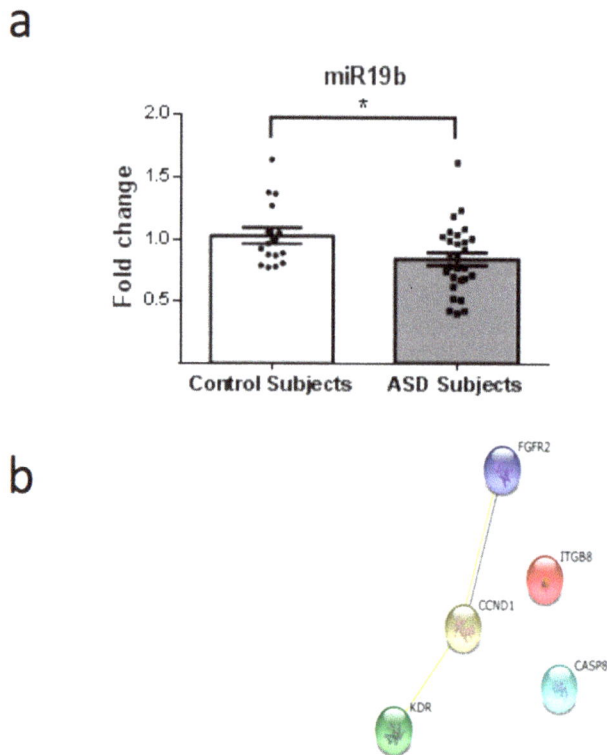

Figure 5. (**a**) Scatter plot of differential relative expression of miR19b in peripheral blood of ASD subjects compared to control subjects. Results expressed as mean ± standard error. *t*-Test analysis; * *p* < 0.05. (**b**) mRNA validates targets for miR19b selected in miRBase and respective proteins clusters that can be involved in ASD. Protein involved in cell cycle: CCND1 (cyclin D1). Proteins associated with cytoskeleton: ITGB8 (integrin, beta 8); KDR (kinase insert domain receptor). Others: CASP8 (caspase 8); FGFR2 (fibroblast growth factor receptor 2).

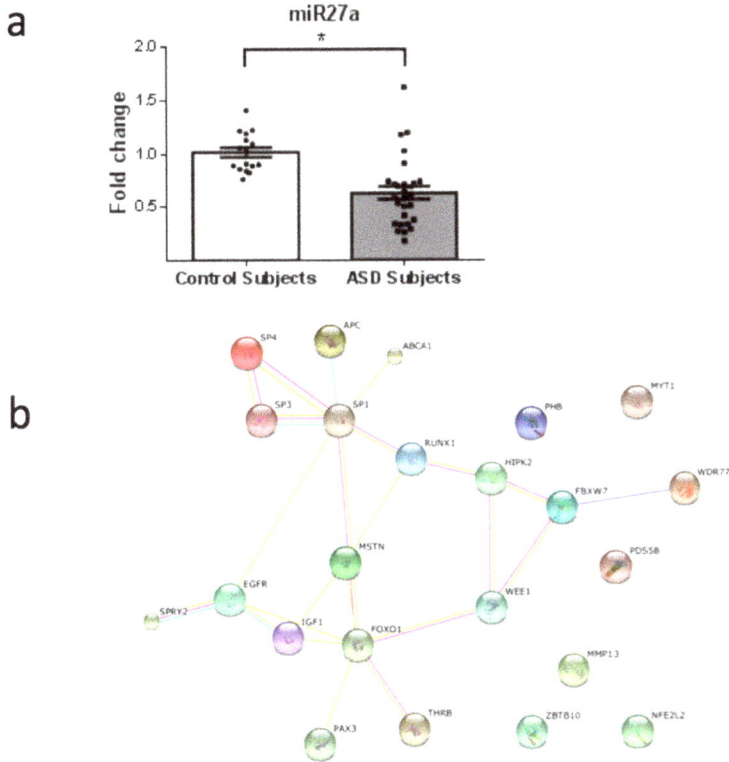

Figure 6. (**a**) Scatter plot of differential relative expression of miR27a in peripheral blood of ASD subjects compared to control subjects. Results expressed as mean ± standard error. *t*-Test analysis; * $p < 0.05$. (**b**) mRNA validates targets for miR27a selected in miRBase and respective proteins clusters that can be involved in ASD. Proteins involved in epigenetic regulation: SP4 (Sp4 transcription factor); SP3 (Sp3 transcription factor); WDR77 (WD repeat domain 77); RUNX1 (runt-related transcription factor 1); MYT1 (myelin transcription factor 1); SP1 (Sp1 transcription factor); FOXO1 (forkhead box O1); PAX3 (paired box 3); NFE2L2 (nuclear factor (erythroid-derived 2)-like 2); HIPK2 (homeodomain interacting protein kinase 2); ZBTB10 (zinc finger and BTB domain containing 10). Proteins involved in cell cycle: PHB (prohibitin), WEE1 (WEE1 homolog). Others: APC (adenomatous polyposis coli), MMP13 (matrix metallopeptidase 13 (collagenase 3)); MSTN (myostatin); EGFR (epidermal growth factor receptor); FBXW7 (F-box and WD repeat domain containing 7); IGF1 (insulin-like growth factor 1); PDS5B (regulator of cohesion maintenance, homolog B); THRB (thyroid hormone receptor, beta); ABCA1 (ATP-binding cassette, sub-family A (ABC1), member 1); SPRY2 (sprouty homolog 2).

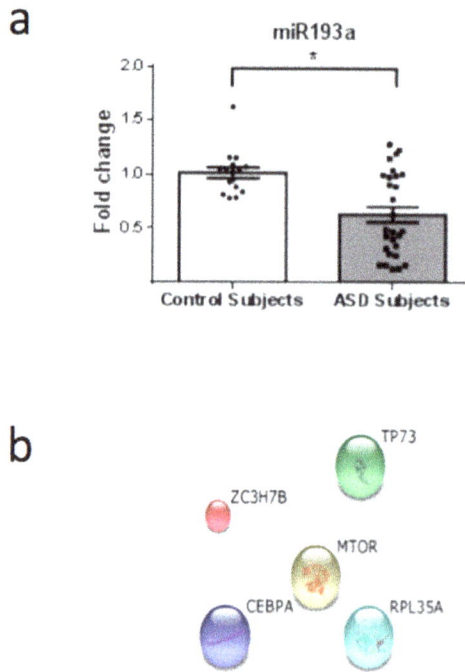

Figure 7. (**a**) Scatter plot of differential relative expression of miR193a in peripheral blood of ASD subjects compared to control subjects. Results expressed as mean ± standard error. *t*-Test analysis; * $p < 0.05$. (**b**) mRNA validates targets for miR193a selected in miRBase and respective proteins clusters that can be involved in ASD. Protein associated with central regulation of cellular metabolism, growth and survival in response to hormones, growth factors, nutrients, energy and stress signals: MTOR (mechanistic target of rapamycin (serine/threonine kinase). Protein involved in cell cycle: TP73 (tumor protein p73). Others: ZC3H7B (zinc finger CCCH-type containing 7B); RPL35A (ribosomal protein L35a); CEBPA (CCAAT/enhancer binding protein (C/EBP), alpha).

Each altered miRNA has a cluster of validated targets, which were predicted for at least two available programs for experimental assays (miRBase). Using the STRING 10 software and information from validated targets for each miRNA available at the miRBase website (mirbase.org), we predicted the pathways that may be involved in ASD.

4. Discussion

ASD is characterized by tremendous phenotypic heterogeneity and the individuals affected demonstrate different levels of behavioral commitment. Considering the differences among ASD individuals, our study aimed to evaluate only one level in the autism spectrum, since the samples tested were only boys that had been clinically diagnosed with classical autism as well established by CARS [19] and (ADOS-G) [20]. As ASD has no biological marker and its etiology is unknown, our research may open new perspectives for investigating cellular processes that occur in affected individuals. The understanding of epigenetic alterations, such as miRNA modifications, could help in elucidate some relevant aspects of ASD, especially in classic autism. Immunological differences could explain the neuro-immunoregulation observed in ASD patients and based in the results shown here, we identified targets that could potentially be affected by altered miRNAs using the miRBase database. The influence of decreased levels of miRNA in ASD patients should be discussed given the fact that the miRNA targets identified are involved in several biological functions that have been previously reported to be dysregulated in ASD, such as cell cycle regulation, axon development and guidance, dendritic spine development and function, protein synthesis regulation and immune response. In addition, it should be taken into account that the intracellular targets for these altered miRNA might also be altered in autistic patients and that this data opens an interesting area of investigation, particularly in blood cell signaling and profiling. Besides the possibility of using this set of miRNA biomarkers for the diagnosis and prognosis of ASD, it is important to understand the pathways that are involved in the regulation of those molecules. Immune aberrations consistent with dysregulated immune responses have been reported in autistic children, including skewed TH1/TH2 cytokine profiles [25], low natural killer (NK) cell activity [26] and an imbalance in serum immunoglobulin levels [27]. In addition, ASD has been linked to autoimmunity and chronic neuroinflammation caused by glutamatergic excitotoxicity and diminished GABAergic signals [28].

The majority of the target proteins of miR34c-5p (Figure 1b)—which is up-regulated in ASD patients—are involved in cell cycle control, such as MYC and MYB. The most important target of miR34c-5p is ZAP70 (Zeta-chain-associated protein kinase 70), a signaling protein involved in immunological development and response. ZAP70 is one of the main proteins involved in lymphocyte activation and function and is also important for NK activity [29]. Additionally, cell-mediated immunity was impaired in ASD patients, as observed by low numbers of CD4$^+$ cells and a concomitant T-cell polarity with an imbalance of Th1/Th2 subsets towards Th2 and the presence of autoantibodies against brain proteins [30]. Inhibition of ZAP70 protein expression could modify the adaptive response and the development of thymocytes. In a study involving 1027 blood samples from autistic children, 45% of a subgroup of children with autism suffered from low NK cell activity [27]. In another study, the cytotoxicity of NK cells was significantly reduced in ASD, as compared with controls [31]. Furthermore, under similar conditions, the presence of perforin, granzyme B, and interferon-gama (IFN-γ) in NK cells from ASD children was significantly lower in comparison to controls [31]. Similarly, it is sustained that lower levels in CD57+CD3− lymphocyte, a subset of NK cells, could be a possible link to a subgroup of ASD [26]. These findings suggest a possible dysfunction of NK cells in children with ASD. Therefore, during critical periods of development, abnormalities in NK cells could represent a susceptibility factor in ASD, that predispose to the development of autoimmunity and/or adverse neuroimmune interactions as already reported [31].

Furthermore, ASD is a neurodevelopmental disorder associated with abnormal neuroplasticity in early development [32]. Another target of miR34c-5p (Figure 1b) is a protein associated with cytoskeleton stabilization in neuronal cell: MAPT (microtubule-associated protein tau with 776 amino acids). The mRNA that encodes MAPT is regulated by alternative splicing, giving rise to several mRNA species [33]. MAPT transcripts are differentially expressed in the nervous system, depending on the stage of neuronal maturation and neuron type [34]. Additionally, mutations in this gene are also associated with neurodegenerative disorders, such as Alzheimer's disease, Pick's disease, frontotemporal dementia, cortico-basal degeneration and progressive supranuclear palsy [35]. Our findings suggest that by epigenetic modifications a negative regulation occurs in the MAPT protein, since changes in neuronal maturation have been widely reported in patients with ASD [33,34]. With regard to epigenetic regulation, the recruitment of histone acetyltransferases (HATs) and histone deacetylases (HDACs) is considered a key element in the dynamic regulation of many genes playing important roles in cellular proliferation and differentiation. The recruitment of HDACs leads to transcriptional repression and inhibitors of this enzymatic activity could reverse aberrant repression and lead to re-expression of genes inducing cell differentiation. In consonance with this hypothesis, the use of Valproic Acid (VPA), a widely used antiepileptic drug, which induces proteasomal degradation of HDAC2 and also inhibits selectively the catalytic activity of class I HDACs [36], during pregnancy is a risk factor associated with the increased incidence of ASD [37]. In this study, we observed that miR145-5p (Figure 2) is up-regulated in ASD patients and could promote mRNA degradation of HDAC2, thus inducing suppression of protein synthesis, since HDAC2 is a validated target of miR145-5p. Therefore, an autistic phenotype could be linked to HDAC2 deregulation due to mir145 up-regulation in our patients.

Considering that SIRT1 enzyme is a NAD$^+$ dependent deacetylase involved in a wide range of cellular processes, it was already shown that SIRT1 negatively regulates the mTOR (mammalian target of rapamycin) signaling, potentially through the TSC1/2 complex [38]. The TSC1/2-mTOR signaling pathway is reported to have a crucial role in mRNA translation during brain development and thus could serve as a crucial pathogenic mechanism in ASD. Furthermore, the up-regulation of miR-199a-5p (Figure 4) may induce a down-regulation of SIRT1 in ASD patients (Figure 8). Additionally, mTOR is a target of miR-193a, another miRNA found to be down-regulated in our study (Figure 7a). TSC1 is a predictive target of miR92a-2a (Figure 3), which was up-regulated in our ASD patients, possibly demonstrating and supporting epigenetic regulation in ASD patients, due to alterations in miRNA in this pathway.

On the other hand, this study demonstrated that the expression of miR-27a-3p (Figure 6) was significantly decreased in ASD patients. Therefore, its mRNA targets could reduce the repression of mRNA translation. An example of a target of miR-27a-3p is insulin-like growth factor-1 (IGF-1), which acts through IGF-1 receptor (IGF-1R, a tyrosine-kinase receptor) [39] and plays a pivotal role in several cellular responses (Figure 8). IGF-1 is a pleiotropic signal in the developing cerebellum, regulating proliferation, neurite outgrowth and survival [40]. The activated IGF-1R activates several phosphorylation cascades being one of the most commonly, the phospho-tidylinositol-3-kinase (PI3K) pathway. PI3K can induce activation of phosphokinase B/Akt (Akt), leading to inhibition of glycogen synthase-kinase-3-beta (GSK3β), and could promote proliferation and survival [41]. The Akt-mTOR pathway is dysregulated in multiple animal models of monogenic causes of ASD, including fragile X mental retardation [42], Rett Syndrome [43] and tuberous sclerosis [44], whereas IGF-1 ligands may improve neurodevelopmental symptoms in Rett Syndrome [45]. In addition, our results show down-regulation of miR-27a (Figure 6) in ASD patients and this could facilitate PI3K activation, indicating that the PI3K/Akt pathway could be up-regulated, which could be involved in the cellular growth, proliferation, migration and adhesion [46].

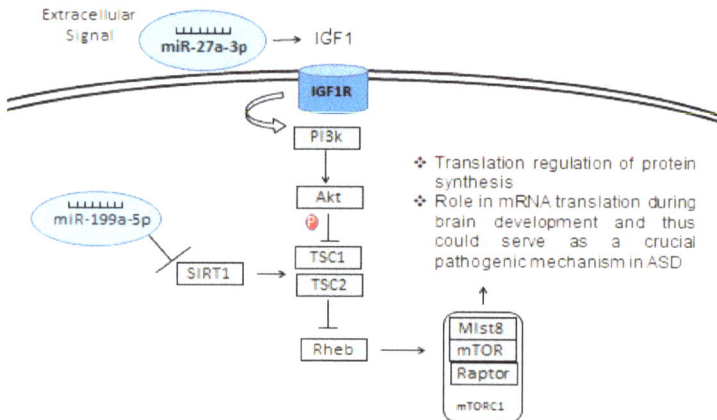

Figure 8. miRNA regulatory pathway in the control of the protein synthesis through SIRT1—TSC:mTOR.

Remarkably, we found that MeCP2 (methyl-CpG binding protein 2), which is involved in Rett Syndrome (RTT), is a target of miR-199a. RTT is an X-linked postnatal neurodevelopmental disorder, which is primarily caused by mutations in the gene encoding MeCP2 [47]. A number of MeCP2 target genes have been identified, including the Brain-Derived Neurotrophic Factor (BDNF) [48]. Several RTT mutations have been described that render MeCP2 incapable of binding to methylated DNA and/or repressing gene transcription. This protein dysfunction has also been shown to cause abnormalities in RNA splicing, suggesting a complex molecular pathogenesis [49]. The loss of the function of the MeCP2 protein leads to autistic behaviors in RTT syndrome. As such, as MeCP2 is a known target of a miRNA found to be altered in our patients with classic autism, we suggest that the MeCP2 protein is decreased by epigenetic inhibition, leading to autistic behavior similar to that observed in patients affected by a mutation in the MeCP2 gene.

5. Conclusions

Despite the intense investigation of the genetic factors involved in ASD, there is still no consensus as to the main causative candidate genes. Therefore, studies of epigenetic regulation could identify altered mechanisms and elucidate the etiology of this disorder. This study shows changes in miRNA levels in patients that could explain many phenotypic features in the spectrum. The investigation of the modulation of proteins that regulate gene expression seems to be a correct path to follow, since the heterogeneity of the disease is an important factor and could be explained by the different levels of regulation. Although the small number of subjects enrolled in our study limits our data, we believe that the results presented here are innovative and important to increase the knowledge in the ASD field. Furthermore, it is important to highlight that the ASD subjects presented the same autism severity, and considering the accuracy and strength of the methodology used, this study produce important hypotheses that should be better explored in future studies. The alterations surveyed in the blood can be seen as permanent changes and correlate from the embryonic development of affected individuals. In addition, the altered miRNA set found in this study may be analyzed in a future molecular diagnosis of the disorder, since the alterations found corroborate with clinical and biochemical situations. The quantification of altered miRNA expression may be used for the prognosis of ASD, as scales of impairment may be made, since this disorder presents a spectrum of severity levels.

Author Contributions: Conceptualization, T.d.S.V.; Methodology, T.d.S.V. and J.M.S.; Software, T.d.S.V. and T.V.; Validation, T.d.S.V., J.M.S. and T.V.; Formal Analysis, T.d.S.V. and T.V.; Investigation, T.d.S.V., J.M.S. and S.S.; Resources, T.d.S.V., J.M.S., S.S., T.V., D.O.S. and R.F.d.A.; Data Curation, T.d.S.V., J.M.S. and S.S.; Writing-Original Draft Preparation, T.d.S.V.; Writing-Review & Editing, T.V., D.O.S. and R.F.d.A.; Visualization, D.O.S.; Supervision, T.V. and R.F.d.A.; Project Administration, D.O.S. and R.F.d.A.; Funding Acquisition, D.O.S.

Funding: This work was supported by CAPES, CNPq and FIPE-HCPA.

Acknowledgments: We would like to thank the individuals who contributed their time and dedication to this study: patients, control children, their parents and the professionals of the Hospital de Clínicas de Porto Alegre, without whom the study would not have been possible.

Conflicts of Interest: Authors declare no actual or potential conflicts of interest. No competing financial interests exist.

References

1. Baio, J.; Wiggins, L.; Christensen, D.L.; Maenner, M.J.; Daniels, J.; Warren, Z.; Kurzius-Spencer, M.; Zahorodny, W.; Robinson Rosenberg, C.; White, T.; et al. Prevalence of Autism Spectrum Disorder Among Children Aged 8 Years—Autism and Developmental Disabilities Monitoring Network, 11 Sites, United States, 2014. *MMWR Surveill. Summ.* **2018**, *67*, 1–23. [CrossRef] [PubMed]
2. Persico, A.M.; Napolioni, V. Autism Genetics. *Behav. Brain Res.* **2013**, *251*, 95–112. [CrossRef] [PubMed]
3. Cristino, A.S.; Williams, S.M.; Hawi, Z.; An, J.Y.; Bellgrove, M.A.; Schwartz, C.E.; Costa Lda, F.; Claudianos, C. Neurodevelopmental and Neuropsychiatric Disorders Represent an Interconnected Molecular System. *Mol. Psychiatry* **2014**, *19*, 294–301. [CrossRef] [PubMed]
4. Glessner, J.T.; Connolly, J.J.; Hakonarson, H. Genome-Wide Association Studies of Autism. *Curr. Behav. Neurosci. Rep.* **2014**, *1*, 234–241. [CrossRef]
5. Meek, S.E.; Lemery-Chalfant, K.; Jahromi, L.B.; Valiente, C. A Review of Gene-environment Correlations and Their Implications for Autism: A Conceptual Model. *Psychol. Rev.* **2013**, *120*, 497–521. [CrossRef] [PubMed]
6. Grafodatskaya, D.; Brian, C.; Szatmari, P.; Weksberg, R. Autism Spectrum Disorders and Epigenetics. *J. Am. Acad. Child Adolesc. Psychiatry* **2010**, *49*, 794–809. [CrossRef] [PubMed]

7. Siniscalco, D.; Cirillo, A.; Bradstreet, A.J.; Antonucci, A. Epigenetic Findings in Autism: New Perspectives for Therapy. *Int. J. Environ. Res. Public Health* **2013**, *10*, 4261–4273. [CrossRef] [PubMed]

8. Huang, F.; Long, Z.; Chen, Z.; Li, J.; Hu, Z.; Qiu, R.; Zhuang, W.; Tang, B.; Xia, K.; Jiang, H. Investigation of Gene Regulatory Networks Associated with Autism Spectrum Disorder Based on MiRNA Expression in China. *PLoS ONE* **2015**, *10*. [CrossRef] [PubMed]

9. Delcuve, G.P.; Mojgan, R.; Davie, J.R. Epigenetic Control. *J. Cell. Physiol.* **2009**, *219*, 243–250. [CrossRef] [PubMed]

10. Abdul, Q.A.; Yu, B.P.; Chung, H.Y.; Jung, H.A.; Choi, J.S. Epigenetic modifications of gene expression by lifestyle and environment. *Arch. Pharm. Res.* **2017**. [CrossRef]

11. Lv, J.; Yongjuan, X.; Wenhao, Z.; Zilong, Q. The Epigenetic Switches for Neural Development and Psychiatric Disorders. *J. Genet. Genom.* **2013**, *40*, 339–346. [CrossRef] [PubMed]

12. Kim, V.N. MicroRNA Biogenesis: Coordinated Cropping and Dicing. *Nat. Rev. Mol. Cell Biol.* **2005**, *6*, 376–385. [CrossRef] [PubMed]

13. Bartel, D.P. MicroRNAs: Genomics, Biogenesis, Mechanism, and Function. *Cell* **2004**, *116*, 281–297. [CrossRef]

14. Kapsimali, M.; Kloosterman, W.P.; de Bruijn, E.; Rosa, F.; Plasterk, R.H.; Wilson, S.W. MicroRNAs Show a Wide Diversity of Expression Profiles in the Developing and Mature Central Nervous System. *Genome Biol.* **2007**, *8*, R173. [CrossRef] [PubMed]

15. Lai, C.Y.; Yu, S.L.; Hsieh, M.H.; Chen, C.H.; Chen, H.Y.; Wen, C.C.; Huang, Y.H.; Hsiao, P.C.; Hsiao, C.K.; Liu, C.M.; et al. MicroRNA Expression Aberration as Potential Peripheral Blood Biomarkers for Schizophrenia. *PLoS ONE* **2011**, *6*. [CrossRef] [PubMed]

16. Zovoilis, A.; Agbemenyah, H.Y.; Agis-Balboa, R.C.; Stilling, R.M.; Edbauer, D.; Rao, P.; Farinelli, L.; Delalle, I.; Schmitt, A.; Falkai, P.; et al. microRNA-34c is a novel target to treat dementias. *EMBO J.* **2011**, *30*, 4299–4308. [CrossRef] [PubMed]

17. Hunt, A.; Shepherd, C. A Prevalence Study of Autism in Tuberous Sclerosis. *J. Autism Dev. Disord.* **1993**, *23*, 323–339. [CrossRef] [PubMed]

18. Creemers, E.E.; Tijsen, A.J.; Pinto, Y.M. Circulating microRNAs: Novel Biomarkers and Extracellular Communicators in Cardiovascular Disease? *Circ. Res.* **2012**, *110*, 483–495. [CrossRef] [PubMed]

19. Schopler, E.; Reichler, R.J.; DeVellis, R.F.; Daly, K. Toward objective classification of childhood autism: Childhood Autism Rating Scale (CARS). *J. Autism Dev. Disord.* **1980**, *10*, 91–103. [CrossRef] [PubMed]

20. Lord, C.; Rutter, M.; Dilavore, P.; Risi, S. *Autism Diagnostic Observation Schedule*; Western Psychological Services: Los Angeles, CA, USA, 1999.

21. Davis, C.J.; Bohnet, S.J.; Meyerson, J.M.; Krueger, J.M. Sleep loss changes microRNA levels in the brain: A possible mechanism for state-dependent translational regulation. *Neurosci. Lett.* **2007**, *422*, 68–73. [CrossRef] [PubMed]

22. Barbato, C.; Ruberti, F.; Cogoni, C. Searching for MIND: microRNAs in neurodegenerative diseases. *J. Biomed. Biotechnol.* **2009**. [CrossRef] [PubMed]

23. Eacker, S.M.; Dawson, T.M.; Dawson, V.L. Understanding microRNAs in neurodegeneration. *Nat. Rev. Neurosci.* **2009**, *10*. [CrossRef]

24. Chen, C.; Ridzon, D.A.; Broomer, A.J.; Zhou, Z.; Lee, D.H.; Nguyen, J.T.; Barbisin, M.; Xu, N.L.; Mahuvakar, V.R.; Andersen, M.R.; et al. Real-Time Quantification of microRNAs by Stem-Loop RT-PCR. *Nucleic Acids Res.* **2005**, *33*, e179. [CrossRef] [PubMed]

25. Gupta, S.; Aggarwal, S.; Rashanravan, B.; Lee, T. Th1- and Th2-like Cytokines in CD4+ and CD8+ T Cells in Autism. *J. Neuroimmunol.* **1998**, *85*, 106–109. [CrossRef]

26. Siniscalco, D.; Mijatovic, T.; Bosmans, E.; Cirillo, A.; Kruzliak, P.; Lombardi, V.C.; Meirleir, K.; Antonucci, N. Decreased Numbers of CD57+CD3− Cells Identify Potential Innate Immune Differences in Patients with Autism Spectrum Disorder. *In Vivo* **2016**, *30*, 83–89. [PubMed]

27. Vojdani, A.; Mumper, E.; Granpeesheh, D.; Mielke, L.; Traver, D.; Bock, K.; Hirani, K.; Neubrander, J.; Woeller, K.N.; O'Hara, N.; et al. Low Natural Killer Cell Cytotoxic Activity in Autism: The Role of Glutathione, IL-2 and IL-15. *J. Neuroimmunol.* **2008**, *205*, 148–154. [CrossRef] [PubMed]

28. Zantomio, D.; Chana, G.; Laskaris, L.; Testa, R.; Everall, I.; Pantelis, C.; Skafidas, E. Convergent Evidence for mGluR5 in Synaptic and Neuroinflammatory Pathways Implicated in ASD. *Neurosci. Biobehav. Rev.* **2015**, *52*, 172–177. [CrossRef] [PubMed]

29. Fischer, A.; Picard, C.; Chemin, K.; Dogniaux, S.; le Deist, F.; Hivroz, C. ZAP70: A Master Regulator of Adaptive Immunity. *Semin. Immunopathol.* **2010**, *32*, 107–116. [CrossRef] [PubMed]

30. Castellani, M.L.; Conti, C.M.; Kempuraj, D.J.; Salini, V.; Vecchiet, J.; Tete, S.; Ciampoli, C.; Conti, F.; Cerulli, G.; Caraffa, A.; et al. Autism and Immunity: Revisited Study. *Int. J. Immunopathol. Pharmacol.* **2009**, *22*, 15–19. [CrossRef] [PubMed]

31. Enstrom, A.M.; Lit, L.; Onore, C.E.; Gregg, J.P.; Hansen, R.L.; Pessah, I.N.; Hertz-Picciotto, I.; Van de Water, J.A.; Sharp, F.R.; Ashwood, P. Altered Gene Expression and Function of Peripheral Blood Natural Killer Cells in Children with Autism. *Brain Behav. Immun.* **2009**, *23*, 124–133. [CrossRef] [PubMed]

32. Kwan, K.Y. Transcriptional Dysregulation of Neocortical Circuit Assembly in ASD. *Int. Rev. Neurobiol.* **2013**, *113*, 167–205. [PubMed]

33. Caillet-Boudin, M.L.; Buée, L.; Sergeant, N.; Lefebvre, B. Regulation of Human MAPT Gene Expression. *Mol. Neurodegener.* **2015**, *10*, 28. [PubMed]

34. Iovino, M.; Agathou, S.; González-Rueda, A.; Del Castillo Velasco-Herrera, M.; Borroni, B.; Alberici, A.; Lynch, T.; O'Dowd, S.; Geti, I.; Gaffney, D.; et al. Early Maturation and Distinct Tau Pathology in Induced Pluripotent Stem Cell-Derived Neurons from Patients with MAPT Mutations. *Brain* **2015**. [CrossRef] [PubMed]

35. Wang, J.Z.; Gao, X.; Wang, Z.H. The Physiology and Pathology of Microtubule-Associated Protein Tau. *Essays Biochem.* **2014**, *56*, 111–123. [CrossRef] [PubMed]

36. Krämer, O.H.; Zhu, P.; Ostendorff, H.P.; Golebiewski, M.; Tiefenbach, J.; Peters, M.A.; Brill, B.; Groner, B.; Bach, I.; Heinzel, T.; et al. The Histone Deacetylase Inhibitor Valproic Acid Selectively Induces Proteasomal Degradation of HDAC2. *EMBO J.* **2003**, *22*, 3411–3420. [CrossRef] [PubMed]

37. Gardener, H.; Spiegelman, D.; Buka, S.L. Prenatal Risk Factors for Autism: Comprehensive Meta-Analysis. *Br. J. Psychiatry* **2009**, *195*, 7–14. [CrossRef] [PubMed]

38. Ghosh, H.S.; McBurney, M.; Robbins, P.D. SIRT1 Negatively Regulates the Mammalian Target of Rapamycin. *PLoS ONE* **2010**, *5*, e9199. [CrossRef] [PubMed]

39. Bondy, C.A.; Cheng, C.M. Signaling by Insulin-like Growth Factor 1 in Brain. *Eur. J. Pharmacol.* **2004**, *490*, 25–31. [CrossRef] [PubMed]

40. Lin, X.; Bulleit, R.F. Insulin-like Growth Factor I (IGF-I) Is a Critical Trophic Factor for Developing Cerebellar Granule Cells. *Brain Res. Dev. Brain Res.* **1997**, *99*, 234–242. [CrossRef]

41. Chin, P.C.; D'Mello, S.R. Survival of Cultured Cerebellar Granule Neurons Can Be Maintained by Akt-Dependent and Akt-Independent Signaling Pathways. *Brain Res. Mol. Brain Res.* **2004**, *127*, 140–145. [CrossRef] [PubMed]

42. Hoeffer, C.A.; Sanchez, E.; Hagerman, R.J.; Mu, Y.; Nguyen, D.V.; Wong, H.; Whelan, A.M.; Zukin, R.S.; Klann, E.; Tassone, F. Altered mTOR Signaling and Enhanced CYFIP2 Expression Levels in Subjects with Fragile X Syndrome. *Genes Brain Behav.* **2012**, *11*, 332–341. [CrossRef] [PubMed]

43. Ricciardi, S.; Boggio, E.M.; Grosso, S.; Lonetti, G.; Forlani, G.; Stefanelli, G.; Calcagno, E.; Morello, N.; Landsberger, N.; Biffo, S.; et al. Reduced AKT/mTOR Signaling and Protein Synthesis Dysregulation in a Rett Syndrome Animal Model. *Hum. Mol. Genet.* **2011**, *20*, 1182–1196. [CrossRef] [PubMed]

44. Pollizzi, K.; Malinowska-Kolodziej, I.; Stumm, M.; Lane, H.; Kwiatkowski, D. Equivalent Benefit of mTORC1 Blockade and Combined PI3K-mTOR Blockade in a Mouse Model of Tuberous Sclerosis. *Mol. Cancer* **2009**, *8*, 38. [CrossRef] [PubMed]

45. Tropea, D.; Giacometti, E.; Wilson, N.R.; Beard, C.; McCurry, C.; Fu, D.D.; Flannery, R.; Jaenisch, R.; Sur, M. Partial Reversal of Rett Syndrome-like Symptoms in MeCP2 Mutant Mice. *Proc. Natl. Acad. Sci. USA* **2009**, *106*, 2029–2034. [CrossRef] [PubMed]

46. Filipowicz, W.; Bhattacharyya, S.N.; Sonenberg, N. Mechanisms of posttranscriptional regulation by microRNAs: Are the answers in sight? *Nat. Rev. Genet.* **2008**, *9*, 102–114. [CrossRef] [PubMed]

47. Díaz de León-Guerrero, S.; Pedraza-Alva, G.; Pérez-Martínez, L. In Sickness and in Health: The Role of Methyl-CpG Binding Protein 2 in the Central Nervous System. *Eur. J. Neurosci.* **2011**, *33*, 1563–1574. [CrossRef] [PubMed]

48. Sun, Y.E.; Wu, H. The Ups and Downs of BDNF in Rett Syndrome. *Neuron* **2006**, *49*, 321–323. [CrossRef] [PubMed]

49. Cheng, T.L.; Zilong, Q. MeCP2: Multifaceted Roles in Gene Regulation and Neural Development. *Neurosci. Bull.* **2014**, *30*, 601–609. [CrossRef] [PubMed]

behavioral sciences

MDPI

Article

Protective Activity of Erythropoyetine in the Cognition of Patients with Parkinson's Disease

Ivonne Pedroso [1,*], Marité Garcia [1], Enrique Casabona [1], Lilia Morales [1], Maria Luisa Bringas [2,*], Leslie Pérez [3], Teresita Rodríguez [3], Ileana Sosa [4], Yordanka Ricardo [1], Arnoldo Padrón [1] and Daniel Amaro [3]

[1]   International Center for Neurological Restoration, La Habana 11300, Cuba; merlincuba@nauta.cu (M.G.); enrique@neuro.ciren.cu (E.C.); lily@neuro.ciren.cu (L.M.); yordanka@neuro.ciren.cu (Y.R.); padron@neuro.ciren.cu (A.P.)
[2]   The Clinical Hospital of Chengdu Brain Science Institute, MOE Key Lab for Neuroinformation, University of Electronic Science and Technology of China, Chengdu 610054, China
[3]   Center of Molecular Immunology, La Habana 11300, Cuba; leslie@cim.sld.cu (L.P.); teresita@cim.sld.cu (T.R.); daniel@cim.sld.cu (D.A.)
[4]   National Center for the Production of Laboratory Animal Breeding, La Habana 10800, Cuba; Iliana.sosa@cenpalab.cu
*   Correspondence: ivon@neuro.ciren.cu (I.P.); maluisabringas@yahoo.com (M.L.B.); Tel.: +53-7273-6777 (ext. 722) (I.P.)

Received: 17 April 2018; Accepted: 18 May 2018; Published: 21 May 2018

Abstract: Introduction: Treatment strategies in Parkinson's disease (PD) can improve a patient's quality of life but cannot stop the progression of PD. We are looking for different alternatives that modify the natural course of the disease and recent research has demonstrated the neuroprotective properties of erythropoietin. In Cuba, the Center for Molecular Immunology (CIM) is a cutting edge scientific center where the recombinant form (EPOrh) and recombinant human erythropoietin with low sialic acid (NeuroEPO) are produced. We performed two clinical trials to evaluate the safety and tolerability of these two drugs in PD patients. In this paper we want to show the positive results of the additional cognitive tests employed, as part of the comprehensive assessment. **Materials and method:** Two studies were conducted in PD patients from the outpatient clinic of CIREN, including $n = 10$ and $n = 26$ patients between 60 and 66 years of age, in stages 1 to 2 of the Hoehn and Yahr Scale. The first study employed recombinant human (rhEPO) and the second an intranasal formulation of neuroEPO. All patients were evaluated with a battery of neuropsychological scales composed to evaluate global cognitive functioning, executive function, and memory. **Results:** The general results in both studies showed a positive response to the cognitive functions in PD patients, who were undergoing pharmacological treatment with respect to the evaluation ($p < 0.05$) before the intervention. **Conclusions:** Erythropoietin has a discrete positive effect on the cognitive functions of patients with Parkinson's disease, which could be interpreted as an effect of the neuroprotective properties of this molecules. To confirm the results another clinical trial phase III with neuroEPO is in progress, also designed to discard any influence of a placebo effect on cognition.

Keywords: cognitive; neuroprotection; Parkinson

1. Introduction

Cognitive symptoms are not the feature that distinguishes Parkinson's disease (PD) but they are very frequent in the natural evolution of this movement disorder [1] and are the cause of many of the disabilities [2]. These appear early [3], even before the onset of the parkinsonian syndrome and progress with it. They are characterized by difficulty in the formation of concepts, in temporal ordering and change of patterns, as well as alterations in attention, motor learning disorders, visuospatial disorders and memory dysfunction all related to the functions of the frontal lobe [4].

The alterations in the executive functions are the fundamental characteristic of the neuropsychological profile [5]. A significant number of patients develop dementia [5–7], which highly impacts their quality of life.

The evolution of cognitive symptoms is variable, from very slow in some cases to rapidly progressing towards dementia in others. The major risk factor is the time of progression and the severity of the motor symptoms, which correlates with the neuropathological stages [8]. Age is another important risk factor [9]. The combination with other non-motor symptoms such as depression and apathy make the patient more vulnerable to the development of dementia [10].

The neural substrate and the pathophysiology of cognitive impairment in PD are not fully known due to their complex nature and the multiple neurotransmission systems that are involved in them [5]. It is known that the pattern of onset of these symptoms is accompanied by alterations of the frontal-subcortical cortex, advancing the lesions towards the posterior cortical zones [7].

In the long term, these symptoms are the most disabling non-motor complications since they impair the mental functions in a permanent way, for up to now there is no way to control its clinical development. The response to the drugs used to treat these symptoms is very low and the progression is unstoppable.

The molecules that are studied for symptomatic treatment, and as neuroprotectors in PD, are mostly evaluated in terms of their action on motor symptoms, not in terms of their action on the neuropsychological sphere [11,12].

In the search for neuroprotective therapeutic alternatives, evidence has been found of the neuroprotective capacity of substances such as erythropoietin, a cytokine known as an important hematopoietic growth factor in tissue oxygenation [13]. It is a glycoprotein hormone that has 165 amino acids, weighs 30.4 kDa and is a member of the cytokine super families. Its location in the adult is the kidney while in the fetus its main production occurs in the liver [14,15].

In 1985 EPOrh was used for the first time in clinical treatments for patients in terminal stages of kidney diseases. In 1989, the US pharmaceutical and food industry approved its use. At present it is widely used in the treatment of anemia related to premature births, renal failure, cancer, chronic inflammatory diseases, and HIV infections [16–18].

Initially it was believed that its only function was to maintain tissue oxygenation at adequate levels; however, it has been shown to have other functions besides being an important hematopoietic factor [19].

The studies showed that EPO has other functions such as neuroprotection, which is performed by several mechanisms which are not all fully clarified. It reduces the toxicity of glutamate, induces the production of antiapoptotic factors, reduces inflammation, decreases the damage mediated by nitric acid, as well as has neurotrophic, antioxidant, and angiogenic action. It also promotes the formation of species reactive to oxygen, the activation of the protein kinase B (PKB) via the kinase-3 phosphoinositide, and the activation of the Janus Kinasa-2 (jak2) and the signaling factor of nuclear factor-kappaB (NF-kB) [19].

EPO has been identified as an important endogenous mediator of the tissue adaptive response to metabolic stress, capable of limiting the extent of tissue damage. In addition, it modifies both the neuronal electrical activity and the synthesis, transport, and release of neurotransmitters in dopaminergic and cholinergic populations, favoring the differentiation of dopaminergic neurons from their precursors under conditions of low oxygen availability in the central nervous system (CNS).

Studies have shown that EPO is capable of modifying the neuronal electrical activity as well as the synthesis, transport, and release of neurotransmitters in dopaminergic and cholinergic populations. Studies carried out in vitro on PC12 cells have demonstrated its ability to stimulate cell survival and increase dopamine release.

EPO is widely distributed in different regions of the body particularly susceptible to aging and atrophy such as the hippocampus, the substantia nigra, and the cerebral cortex. It has been demonstrated that its systemic administration has neuroprotective action in cultures of dopaminergic neurons in animal models MPTP and 6-OHDA of PD, subjected to hypoxia and ischemia induced by the deprivation of oxygen, glucose, glutamate, and excitotoxicity by nitric oxide; however, the striatal levels of catecholamines are maintained in normal limits with the response of a decrease in the rotational asymmetry of the mice [20–22]. In preclinical studies in PD as well as in other neurological [23–25] and psychiatric [26] diseases, EPO has also shown its capacity as a neuroprotector.

The EPOrh obtained in the Center of Molecular Immunology is registered and approved for use in humans in Cuba and other countries. The quality of the product is supported by its increasing international facilitates and its use in therapeutic clinical trials, with a subsequent generalization for use in an open population. In the production process of the EPOrh, isoforms with different sialic acid contents are obtained. When the weight of the EPO molecule is between 4 and 7 mmol/mL of protein, it is considered as having a low content of sialic acid (NeuroEPO). This molecule is similar to that produced in the brain of mammals but does not have an inducer effect in the synthesis of erythrocytes, maintaining its neuroprotective properties.

The name is due to its similarity with the EPO that is synthesized in the brain of mammals and is rapidly degraded by the liver due to its low content of sialic acid. Due to this it must be administered by a non-systemic route, such as intranasal [27,28]. Neuro-EPO is also obtained in the Molecular Immunological Center.

Supported by the knowledge of the characteristics of the molecule obtained through the investigations carried out, as well as the need for medicines and neuroprotective treatment strategies for PD and the economic feasibility of producing the molecule in our country, we carried out clinical studies with the use of this molecule.

We previously conducted a proof of concept clinical trial, administering EPOrh subcutaneously where the neuropsychological performance was measured as a secondary endpoint. Results showed an increased neuropsychological performance of patients after administration as compared to before administration results [29].

In a second investigation led by the author, this time using intranasal neuroEPO, with the main aim to show the safety and tolerability of the drug in PD patients, we took advantage of the neuropsychological evaluation to make a comparison with the first study.

The objective of the current study was to show the most significant results in relation to the neuropsychological scales in the current study performed with neuroEPO and make a comparison with those found previously in the study with EPOrh.

2. Materials and Methods

The first clinical trial was conducted at the International Center for Neurological Restoration CIREN www.ciren.cu and consisted of the administration of EPOrh subcutaneously at a dose of 60 IU/kg, once a week for five consecutive weeks. Ten patients with PD were included according to the Bank of London criteria in stages 1–3 of Hoehn and Yahr. Two patients were in stage 1, one in stage 1.5, one in stage II, five in stage 2.5 and one in stage 3. The age range was between 45 and 67 years with an average of 53 years. The distribution by sex was of 8 men and 2 women.

For the evaluation of cognitive symptoms, the Dementia Rating Scale Scale of Mattis (DRS) [30] was used. Evaluation was performed before starting the treatment, at week 1 of treatment, and 90 days after finishing it.

The statistical analysis was performed with the STATISTICA 6.0 package. The scores resulting from the evaluations were stored in databases.

Descriptive statistics such as the mean and the standard deviation were used for the statistical analysis to know the measurements of the central tendency of the variables and the test of the signs was employed to compare the differences between the scores obtained in the scales before and after the treatment with EPOrh.

All the patients signed an informed consent form as well as their relatives and the clinician conducting the trial.

The second study was conducted by two institutions: The International Center for Neurological Restoration and the Center for Molecular Immunology, who was the promoter. It was a phase I-II physician led clinical trial, in which 26 patients with PD who were classified as being in stages 1-2 according to the Hoehn and Yarh scale participated. The sample was randomly divided into two groups. Group A received neuroEPO and group B received placebo. The group A average age was 56.4 with a 7.8 standard deviation, females predominated at 53.4%, while patients in stage 2 made up 73.4% of the total. The evolution time had a mean of 5.4 years with a standard deviation of 3.2 and most of the patients were college graduates (46%).

In group B the mean age was 61.09 with a 6.6 standard deviation. The male sex prevailed, 72.7%, and stage 2 patients made up 80.8%. The evolution time average was 5.8 years with a standard deviation of 3.5 and most of the patients were college graduates (72.7%).

Despite the apparent differences between the clinical and demographic variables, the evaluation using the Mann-Whitney U and the chi square test did not show significant differences.

Randomization was performed by the CIM assigning N = 15 to group A and N = 11 to group B. The groups were respectively administered neuroEPO and placebo with identical organoleptic characteristics. The informed consent of all patients was obtained before the start of the trial.

The dose of neuroEPO was a vial with a dose of 1 mL/1mg administered intra-nasally for five consecutive weeks. The placebo group was administered 1 mL of an intranasal inert solution for the same period of time.

The assessment of cognitive symptoms was performed before starting the treatment, at week 1 of treatment and six months after its completion, with a wide-range battery of neuropsychological tests. These tests included the Mini mental Examination Test [31], DRS by Mattis, and The Frontal Assessment Battery (FAB) [32] to measure global cognitive functioning. The copy and reproduction of the Rey complex figure [33] was used to evaluate the visuoconstructive function and visual memory considering each one of the 18 units that compose it. The Verbal Fluency Test (D-KEFS) [34], which evaluated dorsolateral frontal lobe functions, consisted of verbal phonological fluency, verbal semantic fluency, and the capacity to alternate mental categories variables. The Word-Color Conflict Test (StroopTest) [35] gave us information about the state of selective and focused attention, the inhibition of responses, and the change of mental set. The Trail Making Test (TMT) [36] assessed the speed of visual location, attention, mental flexibility, working memory, and motor function. The Rey Verbal Auditory Learning Test [37] evaluated functions of the frontal lobe and had as variables working memory capacity, auditory-verbal learning ability, degree of retroactive interference, coding capacity, verbal memory storage, and ability to recognize auditory-verbal information. The working memory index of the WAIS III [38], a multidimensional battery, evaluates the state of the frontal lobe: integrated by subtests of sustained attention, working memory, and problem solving.

The IBM SPSS Statistics V 21 package was used for the statistical analysis of the data. We used tables of frequency analysis and descriptive statistics to analyze the demographic characteristics of the sample. The tables included means and percentages.

For the analysis of the results obtained by studying the differences between quantitative variables for paired and unpaired samples, the Wilcoxon and U of Mann–Whitney tests were used respectively. For the qualitative variables Chi square (X^2) test was used. All values of $p < 0.05$ were considered significant.

3. Results

In the previous study performed with recombinant erythropoietin, administered subcutaneously, the results showed that the cognitive status of all patients measured only by the DRS Scale improved after treatment, their individual score being higher in the final evaluation with respect to the initial one ($z = 2.84$, p: 0.004).

The analysis of cognitive variables in the current study performed with intranasal neuroEPO had the following results:

Phonological and Semantic Verbal Test (FAS):

Subtest the phonological verbal fluency: showed significant differences at week 1 after treatment (z: 2.2, p: 0.02,) in the group treated with neuroEPO, in comparison with the placebo group.

Subtest mental category change: significant differences were found in the group treated with neuroEPO (z: 2.13, $p = 0.03$) at six months after treatment.

Dementia Rating Scale (DRS):

Group treated with neuroEPO: showed significant differences between the pre-treatment evaluation, the week one (p: 0.01, z: 2.5), and six months after the end of treatment (p: 0.005, z: 2.8)

Placebo group: significant differences were found at week one after treatment (p: 0.01, z: 2.5) and at six months (p: 0.011, z: 2.5).

Frontal Assessment Battery (FAB) Test of Litvan:

Group treated with neuroEPO: significant differences in the evaluation made one week after the treatment (p: 0.009, z: 2.5) and six months later (p: 0.004, z: 2.8)

Placebo group: significant differences one week after treatment (p: 0.02, z: 2.3) and six months later (p: 0.009, z: 2.5)

Rey complex figure:

Memory subtest:

Group treated with neuroEPO: significant difference between pre-treatment, at one week (p: 0.0009, z: 3.2) and six months later evaluations (p: 0.001, z: 3.23)

Placebo group: significant differences between the previous evaluation and one week (p: 0.01, z: 2.44) and six months after treatment (p: 0.007, z: 2.66).

Copy subtest:

Group treated with neuroEPO: had a significant difference in the evaluations conducted one week (p: 0.006, z: 2.7) and six months later after treatment (p: 0.007, z: 2.6).

Placebo group: also showed significant difference (p: 0.017, z: 2.3) but only six months after treatment.

The results with other tests, such as the Trail Making Test are not reported in this paper since significant differences were not found.

4. Discussion

In this study we focused on demonstrating some changes in the neuropsychological functions of a group of patients treated with two formulations of the molecule studied.

Our results are preliminary but the authors consider them to be relevant given the importance of cognitive symptoms in PD since they are one of the factors that most affects the quality of life of patients and caregivers, increasing social costs. They do not have a specific treatment and they evolve irreparably [39–41].

Within cognitive disorders in PD, executive alterations are the most important sign whether or not patients develop dementia and they tend to appear early in clinical evolution [42]. They appear accompanied by alterations in visuospatial abilities, spatial orientation, change of mental set, verbal fluidity mainly semantic, initiation, abstraction and generalization of thought, programming of behavior, and some modalities of memory and language [43,44].

Many investigations have shown that patients have symptoms of this type from the early stages of the disease [45].

The verbal fluency test measures the accomplishment of tasks of the set called executive functions that includes actions that require the use of underlying processes of access to the lexicon, and also the ability of cognitive organization, the ability to look for non-habitual words, focal and sustained attention as well as inhibition processes.

The executive functions are also responsible for the anticipation and setting of goals, the formation of plans and programs, the beginning of activities and mental operations, the temporal organization, sequencing, comparison, classification and categorization, the self-regulation of tasks, and the ability to carry them out efficiently [46,47].

These executive functions are processes that are directly linked to the coordinated functioning of the cortical and subcortical systems of the frontal lobes [48,49]. While evaluating them within our study we observed improvement in them, as measured by the verbal Fluency test, in the group exposed to neuroEPO. This was not so in the placebo group that was not exposed to the molecule, in which no significant differences were found in the evaluations after treatment.

In relation to the DRS, FAB of Litvan, and Rey complex figure tests we found that the results were positive for both groups, neuroEPO and placebo, which speaks in favor of the placebo effect.

The definition of the "placebo effect" [50] is well known in the field of research and clinical trials. It is used for the purpose of controlling the psychological effects of treatment.

In clinical research, a placebo is used intentionally to differentiate the pharmacological effects of the study drug from those unrelated to it. In this way it is possible to objectively separate the effects of the studied drug from others produced by the disease or by other factors.

Currently double-blind studies, in which one group of patients receives treatment with the drug under evaluation and the other receives only placebo, is the most adequate choice for the study of new drugs [51–53].

In our study this effect was also observed, for future clinical trials we plan to use tests that are not very susceptible to this effect.

The authors found promising results in the sphere of neuropsychology in PD while investigating neuroprotection. The drug, even administered at low doses because the studies conducted were safety studies, had interesting results, which suggests a possible effect of erythropoietin on the cognitive sphere, as reported already in preclinical studies.

Because this result is very preliminary it is necessary to conduct new studies with an adequate design to demonstrate the possible benefit in cognitive functions. To this aim a Phase II-III clinical trial is being conducted in which we intend to assess whether or not there is a positive impact of the molecule on the symptoms of the disease and its capacity for neuroprotection, given the need for this type of therapy in PD and the necessity to advance in the understanding of the cellular mechanisms of neurodegeneration.

Our studies have several limitations, one of them was that the first study did not have a control group. The other limitation is since both studies were clinical safety trials, small doses of the drug were administered.

However, in spite the fact that the positive impact of EPO on cognition in preclinical and clinical studies has been established in the literature, it has also established that this impact is dose dependent [54]. These findings highlight the need to continue studying the effect of erythropoietin on cognitive functions in PD using higher doses.

On the other hand, the results obtained were based on the total scores of the scales employed, however, we intend in the near future to study the cognitive latent variables related with the drug, which can be revealed using item-response theory, in terms of evaluating the discriminative power of individual items instead of the global scores of the neuropsychological scales.

5. Conclusions

In conclusion, the authors suggest that the beneficial effect in patients undergoing treatment with both EPOrh and neuroEPO could be an effect of the molecules, but since the placebo effect is present, further studies will be necessary to demonstrate the neuropsychological benefits.

Author Contributions: I.P., D.A., T.R., I.S. conceived and designed the study. I.P., M.G., L.M., M.L.B., D.A. analyzed the data. Y.R., A.P., L.P. worked in the studio. E.C. evaluated the protocol patients. I.P. and M.L.B. wrote the paper.

Funding: This research was funded by the Ministry of Public Health of the Republic of Cuba. M.L.B. received funding from the National Science Foundation of China (81330032. 61673090).

Acknowledgments: To Lazaro Alvarez that contribute to increase our knowledge about Parkinson disease.

Conflicts of Interest: The authors declare no conflict of interest.

References

1. Romo-Gutiérrez, D.; Petra-Yescas, M.L.L.; Boll, M.C. Factores genéticos de la demencia en la enfermedad de Parkinson (EP). *Gac. Med. Mex.* **2015**, *151*, 110–118. [PubMed]
2. Litvan, I.; Goldman, J.G.; Tröster, A.I.; Schmand, B.A.; Weintraub, D.; Petersen, R.C.; Mollenhauer, B.; Adler, C.H.; Marder, K.; Williams-Gray, C.H.; et al. Diagnostic Criteria for Mild Cognitive Impairment in Parkinson's Disease: Movement Disorder Society Task Force Guidelines. *Mov. Disord.* **2012**, *27*, 349–356. [CrossRef] [PubMed]
3. Santangelo, G.; Vitale, C.; Trojano, L.; Errico, D.; Amboni, M.; Barbarulo, A.M.; Grossi, D.; Barone, P. Neuropsychological Correlates of Theory of Mind in Patients with Early Parkinson's. *Disease. Mov. Disord.* **2012**, *27*, 98–105. [CrossRef] [PubMed]
4. Janvin, C.C.; Aarsland, D.; Larsen, J.P. Cognitive predictors of dementia in Parkinson's disease: A community-based, 4-year longitudinal study. *J. Geriatr. Psychiatry Neurol.* **2005**, *18*, 149–154. [CrossRef] [PubMed]
5. Pillon, B.; Deweer, B.; Agid, Y.; Dubois, B. Explicit memory in Alzheimer's, Huntington's, and Parkinson's diseases. *Arch. Neurol.* **1993**, *50*, 374–379. [CrossRef] [PubMed]
6. Weintraub, D.; Comella, C.L.; Horn, S. Parkinson's Disease—Part 1: Pathophysiology, Symptoms, Burden, Diagnosis, and Assessment. *Am. J. Manag. Care.* **2008**, *14* (Suppl. S2), S40–S48. [PubMed]
7. Goldman, J.; Weis, H.; Stebbins, G.; Bernard, B.; Goetz, C. Clinical differences among mild cognitive impairment subtypes in Parkinson's disease. *Mov. Disord.* **2012**, *27*, 1129–1136. [CrossRef] [PubMed]
8. Meyer, P.M.; Strecker, K.; Kendziorra, K.; Becker, G.; Hesse, S.; Woelpl, D.; Hensel, A.; Patt, M.; Sorger, D.; Wegner, F.; et al. Reduced $\alpha 4\beta 2^*$–nicotinic acetylcholine receptor binding and its relationship to mild cognitive and depressive symptoms in Parkinson disease. *Arch. Gen. Psychiatry* **2009**, *66*, 866–877. [CrossRef] [PubMed]
9. Williams, C.H.; Mason, S.L.; Evans, J.R.; Foltynie, T.; Brayne, C.; Robbins, T.W.; Barker, R.A. The CamPaIGN study of Parkinson's disease: 10-year outlook in an incident population-based cohort. *J. Neurol. Neurosurg. Psychiatry* **2013**, *84*, 258–264. [CrossRef] [PubMed]
10. Rascol, O. "Disease-modification" trials in Parkinson disease: Target populations, endpoints and study design. *Neurology* **2009**, *72* (Suppl. S2), S51–S58. [CrossRef] [PubMed]
11. Olanow, C.W.; Hauser, R.A.; Jankovic, J.; Langston, W.; Lang, A.; Poewe, W.; Tolosa, E.; Stocchi, F.; Melamed, E.; Eyal, E.; et al. A Randomized, Double-Blind, Placebo-Controlled, Delayed Start Study to Assess Rasagiline as a Disease Modifying Therapy in Parkinson's Disease (The ADAGIO Study): Rationale, Design, and Baseline Characteristics. *Mov. Disord.* **2008**, *23*, 2194–2201. [CrossRef] [PubMed]
12. Zhao, Y.J.; Wee, H.L.; Au, W.L.; Seah, S.H.; Luo, N.; Li, S.C.; Tan, L.C.S. Selegiline use is associated with a slower progression in early Parkinson's disease as evaluated by Hoehn and Yahr Stage transition times. *Parkinsonism Relat. Disord.* **2011**, *17*, 194–197. [CrossRef] [PubMed]
13. Koury, M.J.; Bondurant, M.C. The molecular mechanism of erythropoietin action. *Eur. J. Biochem.* **1992**, *210*, 649–663. [CrossRef] [PubMed]
14. Grasso, G.; Sfacteria, A.; Meli, F.; Passalacqua, M.; Fodale, V.; Buemi, M.; Giambartino, F.; Iacopino, D.G.; Tomasello, F. The role of erythropoietin in neuroprotection: Therapeutic perspectives. *Drug News Perspect* **2007**, *20*, 315–320. [CrossRef] [PubMed]

15. Mainie, P. Is there a role for erythropoietin in neonatal medicine? *Early Hum. Dev.* **2008**, *84*, 525–532. [CrossRef] [PubMed]

16. Halitchi, C.I.; Munteanu, M.; Brumariu, O. Factors influencing responsivenessto treatment in children with renal anemia in end stage renal disease. *Rev. Med. Chir. Soc. Med. Nat. Iasi* **2008**, *112*, 94–99. [PubMed]

17. Badzek, S.; Curic, Z.; Krajina, Z.; Plestina, S.; Golubic-Cepulic, B.; Radman, I. Treatment of cancer-related anemia. *Coll. Antropol.* **2008**, *32*, 615–622. [PubMed]

18. McPherson, R.J.; Juul, S.E. Recent trends in erythropoietin-mediated neuroprotection. *Int. J. Dev. Neurosci.* **2008**, *26*, 103–111. [CrossRef] [PubMed]

19. Sawada, H.; Shimohama, S. MPP+ and glutamate in the degeneration of nigral dopaminergic neurons. *Parkinsonism Relat. Disord.* **1999**, *5*, 209–215. [CrossRef]

20. Erbaş, O.; Çınar, B.P.; Solmaz, V.; Çavuşoğlu, T.; Ateşo, U. The neuroprotective effect of erythropoietin on experimental Parkinson model in rats. *Neuropeptides* **2015**, *49*, 1–5. [CrossRef] [PubMed]

21. Wu, Y.; Shang, Y.; Sun, S.; Liang, H.; Liu, R. Erythropoietin prevents PC12 cells from 1-methyl-4-phenylpyridinium ion induced apoptosis via the Akt/GSK-3b/ caspase-3 mediated signaling pathway. *Apoptosis* **2007**, *12*, 1365–1375. [CrossRef] [PubMed]

22. Dhanushkodi, A.; Akano, E.O.; Roguski, E.E.; Xue, Y.; Rao, S.K.; Matta, S.G.; Rex, T.S.; McDonald, M.P. A single intramuscular injection of rAAV-mediated mutant erythropoietin protects against MPTP-induced parkinsonism. *Genes Brain Behav.* **2013**, *12*, 224–233. [CrossRef] [PubMed]

23. Ehrenreich, H.; Fischer, B.; Norra, C.; Schellenberger, F.; Stender, N.; Stiefel, M.; Sirén, A.L.; Paulus, W.; Nave, K.-A.; Gold, R. Exploring recombinant human erythropoietin in chronic progressive multiple sclerosis. *Brain* **2007**, *130*, 2577–2588. [CrossRef] [PubMed]

24. Boesch, S.; Sturm, B.; Hering, S.; Goldenberg, H.; Poewe, W. Scheiber-Mojdehkar BFriedreich's Ataxia: Clinical Pilot Trial with Recombinant Human Erythropoietin. *Ann. Neurol.* **2007**, *62*, 521–524. [CrossRef] [PubMed]

25. Ehrenreich, H.; Hasselblatt, M.; Dembowski, C.; Cepek, L.; Lewczuk, P.; Stiefel, M.; Rustenbeck, H.H.; Breiter, N.; Jacob, S.; Knerlich, F.; et al. Erythropoietin Therapy for Acute Stroke Is Both Safe and Beneficial. *Mol. Med.* **2002**, *8*, 495–505. [PubMed]

26. Ehrenreich, H.; Hinze-Selch, D.; Stawicki, S.; Aust, C.; Knolle-Veentjer, S.; Wilms, S.; Heinz, G.; Erdag, S.; Jahn, H.; Degner, D.; et al. Improvement of cognitive functions in chronic schizophrenic patients by recombinant human erythropoietin. *Mol. Psychiatry* **2007**, *12*, 206–220. [CrossRef] [PubMed]

27. Lagarto, A.; Bueno, V.; Guerra, I.; Valdés, O.; Couret, M.; López, R.; Vega, Y. Absence of hematological side effects in acute and subacute nasal dosing of erythropoietin with a low content of sialic acid. *Exp. Toxicol. Pathol.* **2011**, *63*, 563–567. [CrossRef] [PubMed]

28. García-Rodríguez, J.C. The Therapeutic Potential of Neuro-EPO Administered Nasally on Acute Cerebrovascular Disease. *Curr. Psychopharmacol.* **2012**, *1*, 1–5.

29. Pedroso, I.; Bringas, M.L.; Aguiar, A.; Morales, L.; Alvarez, M.; Valdés, P.A.; Alvarez, L. Use of Cuban Recombinant Human Erythropoietin in Parkinson's Disease Treatment. *MEDICC Rev.* **2012**, *14*, 11–17. [PubMed]

30. Carvalho, V.A.; Machado, T.H.; Reis, G.C.; Tumas, V.; Caramelli, P.; Nitrini, R.; Porto, C.S. Mattis Dementia Rating Scale (DRS) Normative data for the Brazilian middle-age and elderly populations. *Dement. Neuropsychol.* **2013**, *7*, 374–379.

31. Folstein, M.F.; Folstein, S.E.; McHugh, P.R. Mini mental State. A practical method for grading the cognitive state of patients for the clinician. *J. Psychiatr. Res.* **1975**, *12*, 189–198. [CrossRef]

32. Dubois, B.; Slachevsky, A.; Litvan, I.; Pillon, B. The FAB A frontal assessment battery at bedside. *Neurology* **2000**, *55*, 1621–1626. [CrossRef] [PubMed]

33. Rey, A. *Rey: Test de Copia y Reproduccion de Memoria de Figura Geometrica Compleja*; TEA ediciones: Madrid, Spain, 1997.

34. García, E.; Rodríguez, C.; Martín, R.; Jiménez, J.E.; Díaz, S.H.A. Test de Fluidez Verbal: Datos normativos y desarrollo evolutivo en el alumnado de primaria. *Eur. J. Educ. Psychol.* **2012**, *5*, 53–64. [CrossRef]

35. Martín, R.; Hernández, S.; Rodríguez, C.; García, E.; Díaz, A.; Jiménez, J.E. Datos normativos para el Test de Stroop: Patrón de desarrollo de la inhibición y formas alternativas para su evaluación. *Eur. J. Educ. Psychol.* **2012**, *5*, 39–51. [CrossRef]

36. Reitan, R.M. Validity of the Trail Making Test as an indicator of organic brain damage. *Percept. Mot. Skills.* **1958**, *8*, 271–276. [CrossRef]

37. Litvan, I.; Aarsland, D.; Adler, C.H.; Goldman, J.G.; Kulisevsky, J.; Mollenhauer, B.; Rodriguez-Oroz, M.C.; Tröster, A.I.; Weintraub, D. MDS Task Force on Mild Cognitive Impairment in Parkinson's disease: Critical Review of PD-MCI. *Mov. Disord.* **2011**, *26*, 1814–1824. [CrossRef] [PubMed]

38. Kolb, B.; Whishaw, I. Neuropsychological assesment. In *Fundamental Human Neuropsychology*, 5th ed.; Saunders: Philadelpia, PA, USA, 2002; pp. 751–763.

39. Victoria, M.V.; Ladera, V. Neuropsicología de la Enfermedad de Parkinson. *Rev. Neuropsicol. Neurocienc.* **2012**, *12*, 219–241.

40. Dubois, B.; Pillon, B. Cognitive deficits in Parkinson's disease. *J. Neurol.* **1997**, *244*, 2–8. [CrossRef] [PubMed]

41. Aarsland, D.; Bronnick, K.; Williams-Gray, C.; Weintraub, D.; Marder, K.; Kulisevsky, J.; Burn, D.; Barone, P.; Pagonabarraga, J.; Allcock, L.; et al. Mild cognitive impairment in Parkinson disease: A multicenter pooled analysis. *Neurology* **2010**, *75*, 1062–1069. [CrossRef] [PubMed]

42. Garzón-Giraldo, M.L.D.; Montoya-Arenas, D.A.; Carvajal-Castrillón, J. Perfil clínico y neuropsicológico: Enfermedad de Parkinson/enfermedad por cuerpos de Lewy. *CES Med.* **2015**, *29*, 255–270.

43. Williams-Gray, C.H.; Evans, J.R.; Goris, A.; Foltynie, T.; Ban, M.; Robbins, T.W.; Brayne, C.; Kolachana, B.S.; Weinberger, D.R.; Sawcer, S.J.; et al. The distinct cognitive syndromes of Parkinson's disease: 5 Year follow-up of the CamPaIGN cohort. *Brain* **2009**, *132*, 2958–2969. [CrossRef] [PubMed]

44. Beyer, M.K.; Janvin, C.C.; Larsen, J.P.; Aarsland, D. A magnetic resonance imaging study of patients with Parkinson's disease with mild cognitive impairment and dementia using voxel-based morphometry. *J. Neurol. Neurosurg. Psychiatry* **2007**, *78*, 254–259. [CrossRef] [PubMed]

45. Sawamoto, N.; Piccini, P.; Hotton, G.; Pavese, N.; Thielemans, K.; Brooks, D.J. Cognitive deficits and striato-frontal dopamine release in Parkinson's disease. *Brain* **2008**, *131*, 1294–1302. [CrossRef] [PubMed]

46. Rebollo, M.; Montiel, S. Atención y funciones ejecutivas. *Rev. Neurol.* **2006**, *42*, 53–57.

47. Fernández, M.; Lens, M.; López, A.; Puy, A.; Dias, J.; Sobrido, M. Alteraciones de la esfera emocional y el control de los impulsos en la enfermedad de Parkinson. *Rev. Neurol.* **2010**, *50*, 41–49.

48. Pereira, J.B.; Junqué, C.; Martí, M.J.; Ramirez-Ruiz, B.; Bargalló, N.; Tolosa, E. Neuroanatomical substrate of visuospatial and visuoperceptual impairment in Parkinson's disease. *Mov. Disord.* **2009**, *24*, 1193–1199. [CrossRef] [PubMed]

49. Uc, E.; Rizzo, M.; Anderson, S.; Qian, S.; Rodnitzky, R.; Dawson, J. Visual dysfunction in Parkinson disease without dementia. *Neurology* **2005**, *65*, 1907–1913. [CrossRef] [PubMed]

50. Tempone Pérez, S.G. El placebo en la práctica y en la investigación Clínica. *An. Med. Interna (Madrid)* **2007**, *24*, 249–252. [CrossRef]

51. Íbarra, H.S. El efecto placebo en los ensayos clínicos con antidepresivos. *Acta Bioethica* **2009**, *15*, 172–178.

52. Diederich, N.J.; Goetz, C.G. The placebo treatments in neurosciences New insights from clinical and neuroimaging studies. *Neurology* **2008**, *71*, 677–684. [CrossRef] [PubMed]

53. Požgain, I.; Požgain, Z.; Degmečić, D. Placebo and nocebo effect: A mini-review. *Psychiatr. Danub.* **2014**, *26*, 100–107. [PubMed]

54. García-Rodríguez, J.C.; Teste, I.S. The Nasal Route as a Potential Pathway for delivery of Erythropoietin in the Treatment of Acute Ischemic Stroke in Humans. *Sci. World J.* **2009**, *9*, 970–981. [CrossRef] [PubMed]

MDPI

St. Alban-Anlage 66

4052 Basel

Switzerland

Tel. +41 61 683 77 34

Fax +41 61 302 89 18

www.mdpi.com

Behavioral Sciences Editorial Office

E-mail: behavsci@mdpi.com

www.mdpi.com/journal/behavsci

www.ingramcontent.com/pod-product-compliance
Lightning Source LLC
Chambersburg PA
CBHW051315020426
42333CB00028B/3345